ESSAYS ON THE THEORY OF SCIENTIFIC COGNITION

SYNTHESE LIBRARY

STUDIES IN EPISTEMOLOGY,

LOGIC, METHODOLOGY, AND PHILOSOPHY OF SCIENCE

Managing Editor:

VOLUME 210

JERZY KMITA

*Professor of Logic and Methodology of Science,
Adam Mickiewicz University, Poznań*

ESSAYS
ON THE THEORY OF
SCIENTIFIC COGNITION

Translated from the Polish by Jacek Holówka

KLUWER ACADEMIC PUBLISHERS
DORDRECHT / BOSTON / LONDON

PWN—POLISH SCIENTIFIC PUBLISHERS
WARSZAWA

Library of Congress Cataloging-in-Publication Data

Kmita, Jerzy.
  [Szkice z teorii poznania naukowego. English]
  Essays on the theory of scientific cognition / Jerzy Kmita;
translated from the Polish by Jacek Hołówka.
    p.   cm. — (Studies in epistemology, logic, methodology, and philosophy of science)
(Synthese library; v. 210)
    Translation of: Szkice z teorii poznania naukowego.
    Includes bibliographical references.
    ISBN-13: 978-0-7923-0441-8
    1. Knowledge, Theory of. 2. Methodology. 3. History — Philosophy.
4. Science — Philosophy. 5. Philosophy, Marxist. I. Title.
II. Series. III. Series: Synthese library: v. 210.
BD166.K5713 1990
121 — dc20   89-13485

ISBN-13: 978-94-010-6698-3   e-ISBN-13:978-94-009-0473-6

DOI: 10.1007/978-94-009-0473-6

Published by PWN—Polish Scientific Publishers,
Miodowa 10, 00-251 Warszawa, Poland
in co-edition with Kluwer Academic Publishers,
P.O. Box 17, 3300 AA Dordrecht, The Netherlands

Distributors for the U.S.A. and Canada :
Kluwer Academic Publishers,
101 Philip Drive, Norwell, MA 02061, U.S.A.

Distributors for Albania, Bulgaria, Chinese People's Republic, Cuba,
Czechoslovakia, Hungary, Korean People's Republic, Mongolia, Poland,
Romania, the U.S.S.R., Vietnam and Yugoslavia :
ARS POLONA,
Krakowskie Przedmieście 7, 00-068, Warszawa, Poland

Distributors for all remaining countries :
Kluwer Academic Publishers Group,
P.O. Box 322, 3300 AH Dordrecht, The Netherlands

Translation from the Polish original
Szkice z teorii poznania naukowego,
published in 1976 by Państwowe Wydawnictwo Naukowe, Warszawa

TABLE OF CONTENTS

INTRODUCTION

This collection of essays is, generally speaking, a continuation of the themes that I have discussed in my book *Z metodologicz-nych problemów interpretacji humanistycznej*.[1] Naturally, by continuation I mean a resumption of certain questions that were already dealt with in the previous book. But I treated them rather summarily there, and now I will present them in greater detail. While doing so, I will have to face new issues, which were ignored previously, but which have to be solved, at least tentatively, in order to make it possible for me to elaborate more fully the suggestions made a few years ago. Along with this reason, a second one can be added to explain why in these *Essays on the Theory of Scientific Cognition* I tackle new problems, treated no more than marginally in *Z metodologicznych problemów interpretacji humanistycznej*. The current elaboration of my earlier, incipient suggestions is based on the assumptions which may well serve as a starting point for an investigation of more general questions fundamental for epistemology and frequently formulated within its scope. To make at least a partial use of the opportunity, I concentrated on some of these questions, especially in the later chapters of this book.

As I will often refer in these *Essays* to what has been established in *Z metodologicznych problemów interpretacji humanistycznej* — a fact already mentioned above — in the first part of this Introduction I will present an outline of the main ideas of my former book insofar as the individual papers in this collection may hinge upon them. This will discharge me of the obligation to restate over and over again the tenets I have argued for elsewhere. I must say, however, that while expounding these ideas anew, I felt obliged to make one correction. It is connected with the necessity to eliminate some inconsistencies or understatements that I inadvertently made at the time of writing *Z metodologicznych problemów interpretacji humanistycznej*, and which arose from my opposition methodological individualism. The point of view defined by this opposition will now be called (methodological) anti-individualism.[2] Henceforth I wil refer to my position in its corrected version, pointing out occasionally, in case of a major difference, how my current views vary from those stated originally. I will do so in an effort to impart a sense of unity to this collection of essays and to save their leading motif from excessive digressing.

In *Z metodologicznych problemów interpretacji humanistycznej*, I was preoccupied, first of all, with the investigative procedure mentioned in the title of the book, i.e., humanist interpretation, and secondly, with the affinities that connect this procedure with other types of explanation. I concentrated on humanist interpretation because this kind of explanation is particularly important for the humanities. It is a procedure specifically employed in these disciplines, and besides, due to the manner in which it was characterized. it can be identified and contrasted with several basic orientations in the philosophy of the humanities, such as positivist psychologism, methodological naturalism or anti-naturalism. As I see it, humanist interpretation provides an answer to the question: Why a particular activity has been undertaken by a particular individual? Sometimes this question may be formed: Why the individual in question has endowed the object that he/she has created, with the properties that he/she assigned to it? The answer relies on the general assumption of rationality, which says for every individual (the simplest case being the situation of certainty) that if he/she singles out from the point of view of his/her knowledge a certain set of mutually exclusive but complementary activities, marks out corresponding results of those activities (relevant from the point of view of that individual) and imposes a preferential ordering on them, then he/she chooses the activity which brings about the result of the highest expected preference. Along with the assumption of rationality, the *explanans* of humanist interpretation contains the so-called initial conditions, which characterize through appropriate judgments the knowledge of a concrete individual and his/her preferential ordering, also called the order of values. The selected result, (subjectively) connected with the activity undertaken by the individual, is called the sense of the activity. The concepts used in a detailed analysis of the components of the *explanans* of humanist interpretation, as well as those concepts that can be defined with their help, can further be used to define more concepts, some of which, like sign, cultural system, culture, etc., are needed in the explication of the terms employed in all humanistic disciplines. Still other, such as the concepts of the representing structure, the represented structure, the message conveyed by a work of art, reductive (realistic) type of communication, deductive (symbolic-metaphoric) communication, etc., make it possible to explicate the terms used in specific branches of the humanities, especially in the science of arts.

In point of fact, the two central topics in Z *metodologicznych problemów interpretacji humanistycznej*, namely, the problem of the methodological properties of humanist interpretation and the problem of its relation to other explanatory procedures, are closely connected. The following remarks will substantiate this claim.

One of the most important methodological questions that concern humanist interpretation reads as follows: Since the premises of an explanation make account of the beliefs of a given, concrete individual in an idealizing mode (as in point of fact, the subject is not fully aware of the beliefs ascribed to him/her in the interpretation, hardly ever follows the consequences which are ascribed to him/her on the strength of the knowledge and language that he/she uses) and the activities undertaken by the subject (widely) differ from those expected

from him/her on the basis of the premises used in the interpretation, how can we control the adequacy of these premises,
and what does the word 'adequacy' mean in this situation?
With this another question is connected: Putting aside the
idealizing character of humanist interpretation, how can one
guarantee that it will not degenerate into an *ad hoc* explanation,
i.e., how can one make sure that it is supported by something
above the mere fact that the subject under interpretation has
undertaken certain activities?

Both problems can be solved by a recourse to the same set
of assumptions. An act of humanist interpretation is in fact
an act of ascribing to a given individual a system of beliefs
that constitutes a part of the *social consciousness* of a group to
which that individual belongs. The actual convictions of the
individual may deviate more or less from that system. Hence,
in an interpretation which ascribes to an individual an assent
to a certain system of views, e.g., by assuming that a concrete
speaker or hearer of a linguistic utterance possesses linguistic
competence in the sense used by N. Chomsky, we perform an
idealization. A confrontation between the activity undertaken
by an individual, as it has been perceived from the point of
view of an idealizing humanist interpretation, and the activity
that has been actually performed, reveals discrepancies that
must be explained by taking into account the difference between social consciousness and actual convictions of the individual in question. An explanation of a given activity in terms
of the differences between the contents of individual consciousness and the contents of social consciousness is called by
me concretized humanist interpretation.[3]

The question in what conditions an idealizing humanist interpretation does not deserve disqualification as an *ad hoc*
explanation, and also the question in what conditions an interpretation will be considered adequate, can both be answered
by relying on the concept of social consciousness.[4] One performs an idealizing humanist interpretation of the fact that
a certain action has been undertaken if one explains this fact

by the assumption of rationality of the agent, and by relating the point of view of the agent to an appropriate fragment of social consciousness. Thus these two questions reduce ultimately to the problem how social consciousness can be theoretically reconstructed. In other words, humanist interpretation depends on a more fundamental theoretical explanation which pertains to the contents of social consciousness.

In Z metodologicznych ..., I have made two assumptions concerning this reduction. First, that the fundamental research procedure for humanist interpretation, and, a fortiori, a fundamental procedure for all forms of humanist investigation, is genetic explanation (of the transformations of social consciousness), and second, that the primarily task is to be performed by functional-genetic explanation, which '(...) represented (...) above all by Marxism has the prerequisities which recommend it for the principal role in the venture that we have in mind.' [5]

These two assumptions and the questions connected with them which concern the characteristics and the theoretical-and-methodological context of functional-genetic explanation are the point of departure of these Essays. As to the conception of functional-genetic explanation, it cannot be left in the form which I gave it in Z metodologicznych ..., but has to be corrected to some extent, as will be done below. I will present it now in a brief outline, since directly or indirectly this entire book is devoted to this form of explanation.

Functional-genetic explanation must formulate the sufficient condition for the occurrence of either of the two kinds of situation: (1) that a certain set of beliefs remains in social consciousness, (2) that a new set of beliefs emerges in social consciousness. In both cases, the sufficient condition comprises two contributive components which explain the phenomenon in question: (1) the objective function of the relevant set of beliefs, i.e., some sort of objective demand that the beliefs satisfy, (2) the amount of the pertinent, antecedently

accumulated 'thought material', in Engels' sense of the word.
Insofar, as the 'thought material' is functional, it remains in
social consciousness, whereas new beliefs, individually articulated, have a good chance of penetrating into the realm
of social consciousness [6] and may introduce a change therein
step by step as this change makes social consciousness more
functional, and especially in proportion to the deficiency of
functionality of social consciousness as long as it is deprived of
this modification.

The conception of functional-genetic determination is offered to solve the problem of the circumstances sufficient to
open social consciousness to a change and to let in certain
individual beliefs. A question may be asked in this context
that, at first blush, seems absurd, especially to those whose
beliefs have been shaped by methodological individualism: Is
it so that all beliefs which constitute the realm of social
consciousness must have been filtered at one time or another
by somebody's individual consciousness? or: Is it necessary
that before a belief settles in social consciousness it must be
articulated and accepted by an individual? This question only
appears to be absurd, and a brief reflection is enough to realize
that there are such beliefs which are, in a sense, socially
accepted, or (in a weaker version) respected, even though
no one has articulated them individually, let alone accepted
them. The fact was clear even to K. R. Popper, who represents
the position of methodological individualism in such version
which, being much more refined than the naive, psychologistic,
methodological individualism of positivists, is not vitiated by
his own conception of 'the third world', a specific counterpart
of social consciousness. Thus Popper allows for the existence
of the very kind of beliefs that we are concerned here with,
such as unconscious consequences (i.e., not recognized consciously by any individual) of known and accepted truths.
In fact, one can adduce more dramatic examples than that.
For instance, we all respect language rules which make up our
social linguistic competence, although we can neither enu-

merate these rules nor point to an individual who 'invented' them in the past. Moreover, there is little likelihood that a commonly shared belief, e.g., that the sunset is followed by the onset of darkness, although articulated and accepted uniformly by virtually everybody, and therefore part of social consciousness, could have gained this status had it not first been at some time articulated and accepted by one individual.

These problems are crucial for the conception of functional-genetic determination of social consciousness. If it could be shown that any belief, or better a certain class of beliefs, may belong to the realm of social consciousness without first being articulated by individuals, then the idea of functional-genetic mechanism of change in social consciousness would have to be modified, or at least complemented with a conception of another explanation which would provide a sufficient condition for the occurrence of the contents in social consciousness which have not been produced by concrete individuals. Although it is not my intention to solve this complicated problem in this book, I would like to offer my tentative answer. It seems to me that intuitively most convincing is the view that some beliefs, but only those which are of this special kind, can appear in the realm of social consciousness without intervention of individuals. Consequently, although it is still possible to offer functional-genetic explanation of changes in social consciousness when one faces such cases, strictly speaking, some of these changes at least should be analysed with the help of a different explanatory procedure.

I cannot be held up by this problem, however, because, as I have announced, I will not deal with it in this book. That means that I will make no statement about the right explanatory procedure which should complement functional-genetic explanation. I will only disclose, by way of suggestion, the nature of the beliefs which are present in social consciousness, even though these beliefs are neither consequences of socially truths nor have been antecedently perceived by any individuals. Before I do, however, I must make some remarks concerning the concept of social consciousness. While so doing, I will inevitably preempt some of the ideas to which the essays that follow have been dedicated. I am forced to proceed in this manner as I must not venture to elaborate on the suggestion announced above against the backdrop of two assumptions only : that social consciousness is an idealization of the class of beliefs consciously espoused by concrete individuals, and that in some circumstances the Popperian conception of 'the third world' makes an adequate account of some properties of social consciousness.

In my opinion, the fundamental notion to be used here is that of *social practice*. It will be discussed more extensively by me in the latter part of this collection of essays. Social practice as such breaks down into a number of hierarchical, functionally connected types. This differentiation and the system of functional connections constituting a network of different types of social practices are determined by the stage of historical development, notwithstanding the fact that the network is also characterized by certain invariants, such as the functional dominance of 'material element', i.e., productive practice. Every type of social practice can be identified with an autonomous segment of social division of labour (in the broad sense of the term 'labour'). It contains a certain subjective-social regulator which has the form of a system of beliefs, respected by individuals, and which singles out certain values as objectives to be achieved, together with methods of their achievement. As a rule, these beliefs can be verbalized, an obvious fact in the light of the foregoing remarks, and formulated as normative utterances (identifying values) and directives (identifying the methods to achieve these values). This subjective-social regulator of social practice of a particular type is subordinated to an objective demand.[7] This means that the activity selected by the regulator satisfies the condition stipulating that in the subjective-rational sense it contributes to the creation of an objective additive effect which responds to, and fulfils, the demand. Demand can be viewed as a potential objective effect of a certain set of possible actions. (The meaning of the word 'possible', as used here, will be elucidated further down in these *Essays*). This effect is to be so constructed that without it the activities undertaken previously on social scale would gradually lose their efficiency. It can be further mentioned that the fulfillment of a given objective demand results from the performance of such activities which, by bringing about an additive objective effect, make it possible to achieve historically determined and socially accepted values.

This statement characterizes the concept of objective demand only 'superficially'. Effective implementation of historically created and socially accepted values is only a symptom of the fact that social demands have met with an adequate response. In this Introduction, I will present elements of theoretically more fundamental characteristics of this concept.

An example of what I call the subjective-social regulator of a given type of social practice, or to put it in different terms, an example of a particular form of social consciousness is linguistic competence. It is a subjective-social communication regulator of linguistic practice and consists of such specific norms and directives which satisfy an objective demand. This demand is functionally subordinated to further demands of other types of social practice, in first place to 'material practice'. The demands of linguistic practice are thus secondary ones.[8]

Social consciousness, which I will identify with a set of forms of social consciousness corresponding to particular types of social practice, may also be characterized in a different way; for instance, from the point of view which distinguishes between the contents which are and those which are not functionally subordinated to social class demands or to demands of society as a whole. Obviously, the effects of this differentation will vary from one socio-economic formation to another, or even between different stages or historical versions of the same formation. I am neither particularly concerned here with different functions connected with different historical local-

izations of the same sets of beliefs, nor with historically different functions of the same spheres of social consciousness.
I am now talking about a special kind of differentiation of
social consciousness derived from the distinction between
social practical knowledge and its theoretical-scientific or
Weltanschauung complement. A set-theoretical union of sets
of beliefs which make up the last two spheres of social consciousness will be called the complement of social practical
knowledge. I must emphasize that I do not treat the components of this union as mutually exclusive, on the contrary,
I believe that the set-theoretical products of their successive
historical 'incarnations' are not empty.

Under social practical knowledge, I understand that part of
social consciousness which is composed of general beliefs and
their consequences, i.e., singular logical conditionals super-
venient on them. They represent direct-practical knowledge
with prospective purport to be used by individual participants
in social practice.[9] There is one view of the nature of beliefs
that antecendently do not necessary have to be articulated and
accepted by concrete individuals, which I find intuitively correct, namely, that this class of beliefs is part of the contents
of social practical knowledge. The beliefs which belong to the
complement of social practical knowledge, i.e., to a scientific
theory interpretation or to a *Weltanschauung*, have to be
put in words by concrete individuals before they are socially
accepted. Thus, the belief, e.g., that the sunset is followed by
the onset of darkness, or other elements of the contents of
social practical knowledge, arise spontaneously, so to speak,
from the objective conditions in which all the individuals
participating in a social practice find themselves. They arise
from these conditions and from the 'thought material', i.e.,
mainly from the specific concepts which the individuals in
question have at their disposal. But Lutheranism, for instance,
or the theory of relativity, could be accepted by, respectively,
the consciousness of a specific social group or the community
of physicists only after a prior intellectual activity of concrete

individuals had been performed and culminated in the articulation of these beliefs.

*   *   *

As I have said already, the assumption of rationality is a nomothetic component of humanist interpretation that refers to an
appropriately selected fragment of social consciousness. This
assumption can carry the burden it has been charged with
because, as a matter of fact, beliefs of concrete individuals
cluster round the point represented by social consciousness,
which serves as their idealization. In other words, the postulated methodological role of the assumption of rationality is
based on the regularity arising from the fact that individuals
have to accept, or at least take into account, contents of social
consciousness insofar as they have preferences about values
and a desire to carry them out effecively. If they fail to take
notice of the contents of that consciousness, their activities
will be ineffectual. They will be undertaken in the name of
values that cannot be put into practice at all, or with the view
to implementation of such values which, although practically
acceptable, cannot be achieved by a recourse to knowledge
which has not been accepted. Thus, for instance, various individual values connected with the meanings of individually
undertaken communicative-linguistic utterances require for
their fulfillment that the norms and rules of the linguistic
competence are observed, i.e., they presuppose a linguistic
social consciousness. Various individual values that are to be
materialized through individual practical activity of this or
that capitalist owner of means of production can be materialized only if, among other things, he observes the principle
of maximization of profit. But an individual can also abandon
any desire to implement his/her own values to some extent,
even if his/her survival be at stake, and then he/she becomes
less sensitive to the pressure of social consciousness.

This interpretation of the relationship between individual
consciousness and social consciousness, based on the assump-

tion of rationality of individuals, has at least two important aspects. First, individual consciousness, as defined here, is not conceived as a set of beliefs contained in a wider (naturally) set of beliefs constituting the realm of social consciousness. Secondly, it is not conceived after A. Hauser [10] as a set-theoretical union of two mutually exclusive sets of beliefs, one contained in social consciousness, the other being purely individual or private (and therefore susceptible only to psychological explanation). Instead, it has been assumed that every set of individual beliefs is a specific deformation of an appropriate set of beliefs from the realm of social consciousness; thus, the former remains in such relation to the latter as real gas in a given temperature to the ideal gas, or a concrete pendulum to the mathematical pendulum. Every set of beliefs resides at a certain distance from its ideal social counterpart, and the distance is, naturally, different for different individuals; it is manifested in different ways. The beliefs of an individual may vary as to the degree of deductive arrangement within the system they constitute, the system may be more or less deductively complete, and the contents of social consciousness may themselves bear traces of the circumstances in which they were forged from individually articulated, distorted beliefs.

In the position outlined heretofore, the concept of social consciousness is theoretically more fundamental than the concept of individual consciousness. This means that the fact that certain properties belong to elements of social consciousness, or that these remain in a certain relation to other elements, does not, as a rule, require for the purposes of explanation a recourse to facts about individual consciousness, while the latter has to be explained in terms of social consciousness, even though, normally, some additional assumptions must be made as well. But in general, a set of individual beliefs can only be explained in terms which presuppose a specific characteristic of a related set of beliefs in social consciousness and highlight the discrepancy of the former from its ideal social counterpart.

These assumptions help to identify that version of methodological anti-individualism which has been presupposed in the *Essays*, at least with respect to the relationship between social consciousness and individual consciousness. Methodological individualism finds it possible (or in the normative version, postulates) to reduce phenomena concerning social consciousness to phenomena characterizing individual consciousness in the explanation of the former. Methodological anti-individualism declares, in a weaker or stronger form, that such explanation is impossible. We can distinguish between (i) moderate anti-individualism, (ii) radical anti-individualism and (iii) extreme anti-individualism. According to (i), only some phenomena about social consciousness cannot be reduced in the process of explanation to phenomena concerning facts about individual consciousness. According to (ii) and (iii), no state of affairs assigned to social consciousness can be explained exclusively in terms of phenomena concerning individual consciousness. The difference between (ii) and (iii) is this. Radical methodological anti-individualism recognizes the necessity to refer to facts about social consciousness in the explanation of facts about individual consciousness, but at the same time emphasizes that it is necessary to rely on other explanatory premises as well to explain the cases of difference between an individual consciousness and its ideal social counterpart. Extreme methodological anti-individualism either assumes that the set of beliefs of every individual is contained in social consciousness or, which is more consistent, regards research on individual consciousness as erroneous ('unscientific' or entirely devoid of cognitive content), and consequently rejects the concept of individual consciousness as fallacious.

It can be easily perceived that my position should be called radical methodological anti-individualism. It is incompatible with moderate anti-individualism insofar as I believe, and have mentioned it above, that a theoretical explanation of the phenomena belonging to an individual consciousness is to be provided in terms of phenomena concerning social consciousness. My position is also incompatible with extreme anti-individualism in view of the fact that in my opinion theoretical explanation requires acceptance of some additional explanatory premises, and generally, because extreme anti-individualism is incompatible with functional-genetic conception of explanation of changes which occur in the realm of social consciousness and the assumption that some contents originally conceived in individual consciousness pass to social consciousness. Extreme anti-individualism does not take such contents into account at all. As I see it, an objective demand for a certain kind of beliefs is crucial—and will be satisfied sooner or later, in accordance with the genetic-functional conception—but an intervention of individual intellectual activity is nevertheless indispensable and plays an important role in precipitating an imminent change in social consciousness. Moreover, the concrete form in which the objective demand is satisfied depends on individual intervention.

Obviously, an individual can possibly create new ideas, or, in general, on the view adopted here, creativity, in the broad sense of the term, is possible, insofar as the relationship between individual consciousness and social consciousness is based on the assumption of rationality of individual consciousness. Due to this assumption, various innovations are possible and they can produce individualistic 'deformations' in social consciousness. Then some of these changes may turn out to be creative, in the narrow sense of the term, if they respond with sufficient adequacy to the new objective demands. If they do, they pass into the realm of social consciousness.[11]

* * *

I would like to add some remarks about the functional component of the functional-genetic determination of social consciousness. It can be characterized more precisely with the use of the concept of diachronic-functional structure with respect to a given, global developmental quality. I introduced this concept in *Z Metodologicznych* ... Speaking generally, the structure is a relational system which is predetermined to possess a given, global developmental property (being the property which belongs to the system by virtue of its incessant sustaining a developmental process of a specific kind). The states of the elements of that structure, and of sets of elements, are determined (which is the very functional determination we are trying describe) in such a manner that they let in only such states of elements which are compatible with the further retention of the given global property of the structure. In other words, for every element, or set of elements, there exists at any period of time a repertoire of states, and it is a necessary condition safequarding retention of the aforementioned global property by the structure that every actual state of the elements of the structure is a state from that repertoire. It is also presupposed that the global property never ceases to characterize the structure. In particular, a (diachronic-) functional structure may have a hierarchical character,[12] which means that it breaks down into a number of functional substructures having such global properties which are functionally subordinated to the global properties of appropriate superior substructures, and ultimately, through a number of intermediate stages, to the global property of the entire structure. This means, among other things, that the subordinated functional global properties constitute the underpinning of the superior functional global property.[13]

Social practice, together with its concomitant objective conditions, is treated by me as a special case of a diachronic hierarchical, functional structure due to its (developmental) global property of reproduction of the existing conditions connected with the creation of new conditions of the same kind. It is a hierarchical structure because every type of social practice is a separate, relatively autonomous functional substructure. The structure also has, as I have mentioned before, a functionally superior practice, which dominates all other practices, viz., the 'material' practice. When I say that there is an objective demand for a specific overall effect of a practice of a given type, I mean that this practice is a functional structure with respect to the global property of producing the effect of the type in question; and this property is functionally subordinated to an appropriate global property of the superior (sub)structure, which cannot exist without the former.

As the global developmental property of a diachronic structure constituted by social practice conceived as a whole, together with its concomitant objective conditions, is the property of the reproduction of these conditions and creation of new conditions of the same type, we can say that all objective demands (whether more or less predetermined) are functionally subordinated to the global property in question. The further a given demand is 'removed' from the global developmental property (due to the intervention of a large number of intermediate links), the weaker are social consequences of leaving the demand unsatisfied, until the unfulfilled demands concur in producing a widespread inefficiency of activities undertaken with the view to achieving specific, socially accepted values. A typical example of this situation are the activities that constitute a practice which creates a demand for a functionally subordinated type of social practice.

I consider it of fundamental theoretical importance that the concept of objective demand has the following characteristics. It is an overall (potential) effect of an objective social practice of a given type — with the proviso that the achievement of this effect is a global property of the substructure representing the practice of a given type — which is in turn (and in the last resort) functionally subordinated to the reproduction of the objective conditions of the overall social practice and to the creation of new conditions of the same kind.

I must caution the reader that in *Essays* I will use the term 'objective demand' in two related but distinguishable senses, and that their conflation may cause misunderstanding. In the first sense, which has partly been explicated above, an objective demand that must be satisfied by a given type of social practice is, speaking generally, a certain overall (potential) objective effect of the practice of the type in question. In the second sense, which will be easily recognized by the consistent employment of the phrase 'an objective demand for...', the term denotes the relation of functional superiority between a certain functionally superior objective demand in the first sense of the term (i.e., a demand which is satisfied by a functionally superior type of practice) and a specific, overall (potential) effect of an objective practice of the type in question. Due to the latter relation, this effect acquires the status of objective demand in the first sense of the term. Or to put it in a nutshell, the phrase, 'an objective demand for...' refers to the relation of functional superiority concerning objective demands in the first sense of the term. The expression symbolized by dots denotes clearly and unequivocally the second element of the relation. In will continue using the phrase 'objective demand' without pointing out which sense I have in mind, as the context in which the phrase is embedded will make it sufficiently clear what I mean.

Every functional substructure representing a separate type of practice is characterized by the fact that its further substructure is a subjective-social regulator (or a particular form of social consciousness), whose contents correspond to objective demands created by an appropriate type of social practice and addressed to the subjective-social regulator. The concrete manner in which the demand will be satisfied, i.e., the concrete form of the beliefs that penetrate into the sphere of social consciousness in response to the demand that has been created, is further specified by the available 'thought material' connected with that type of social practice. It does not mean, of course, that their ultimate form is unequivocally predetermined. When we leave aside the case of social practical knowledge (in conformity with the remarks I have made earlier about it) as a plausible exception, the beliefs that compose social consciousness acquire their concrete formulation in individual acts of verbalization.[14]

The foregoing review has presented my basic intuitions which serve as the starting point for the essays that follow. Some of these intuitions will be further discussed, some will be only relied upon. I will proceed now to give an overview of the contents of this book.

The first essay — 'On Two Kinds of Explanation' — characterizes this type of explanation which is representative for functional-genetic explanation and, as it will become clear further down, also for humanist interpretation. It will be presented as a specific trait of historical scences as opposed to mathematically oriented natural sciences. Historical sciences so conceived may be described as theoretical disciplines which focus on developmental processes, and developmental social processes in particular. This formulation indicates that the division of empirical sciences into the humanities and natural sciences (i.e., such that do, or respectively, do not use humanist interpretation and make account for the rationality-*cum*-consciousness component in the phenomena under study), although quite legitimate from the methodological point of view, is not as fundamental as the division into mathematics-*cum*-natural-sciences on the one hand, and historical sciences on the other. The distinction is especially important from the Marxist point of view, which acquired its present form under the influence of 'the discovery of the continent of history', according to a well known phrase of Althusser. This appelation is well conceived, I believe, if we interpret it as the tenet that Marxist theory of social cognition is a response to a historical demand (in the sense explicated above) for a research practice inaugurated to a great extent by the founding fathers of historical materialism.

The next essay — 'On the Concepts of Historical Possibility and Necessity' — makes use of some findings which have been arrived at in the first essay and provides a further analysis of historical explanation. It revolves round the problem of explanation of the social developmental process, conceived as more or less transparent fulfillment of the historical necessities which determine social practice.

The fact that from the point of view of historical materialism science, in one of the meanings of the term, is a specific type of social practice, and, in another meaning, is a form of social consciousness, makes it possible to profit from the findings made in the previous essay, which brings into the open Marxist research assumptions concerning the development of social practice. They can be applied to some hotly debated problems in contemporary philosophy of science and to the problem of the growth of science in particular. These questions have been taken up in the essay on 'The Controversy about the Determinants of the Growth of Science'. It is pointed out in this essay that, on the one hand, Marxist theory of scientific cognition is a historical discipline and its core is a research on the growth of science, and, on the other hand, Marxism rests on assumptions which are in firm opposition to those that have been accepted by adherents of Popper and Kuhn or those of Feyerabend in the debate on this matter. In the essay 'The Controversy...', a transition is made from problems of theory of historical cognition to problems of historical theory of cognition. The 'transition' is so much more natural that Marxist theory of scientific cognition is, as I have mentioned, a historical discipline, and therefore all findings about historical research are true for it as well.

From this perspective, i.e., from the perspective of the science of cognition built upon the foundations of historical materialism, the concept of practical knowledge is characterized in the next essay — 'The Relation of Marxist Epistemology to Empircism'. The essay focuses on the connection between reliable empirical knowledge and practical knowledge. This characteristic is made in opposition to traditional empirism in its three versions: classical positivistic, logical positivistic and hypothetistic.[15]

The last essay is primarily devoted to the following question. My earlier findings concerning the growth of science (conceived either as a social practice of some kind or as a form of social consciousness correlated with that practice) and its connection with practical knowledge have been based on the assumption that social research practice functions objectively (in relation to other types of social practice) in a manner that can be schematically characterized, and which differs from one historical period to another. It turned out that it is possible to present this schematic functioning as a framework for the concept of science (again, either as a practice or as a form of consciousness) which is valid for all historical periods, and from which it is possible to deduce those findings already aluded to. In particular, it can be shown in this way that new sets of research results (theories) presuppose *correspondence* to formerly existing sets of similar results (earlier theories) before they win social acceptance. Correspondence is therefore a method of leaning on the 'accumulated thought material', specific for research practice. Consequently, the question arises: What are other methods of establishing ties with the 'accumulated thought material', characteristic for other types of social practice, and how do they differ from correspondence? The last essay — 'Adaptation as the Opposition to Correspondence' — is concerned with a comparison of correspondence with that form of establishing a continuity with tradition which is characteristic for social artistic practice. This method — which relies on a framework in which the objective function of the artistic practice is presented (not unlike the framework of research practice) — has been called *adaptation.*

When I offer the comparison of adaptation and correspondence, I want to kill two birds with one stone. First, I want to highlight the peculiar features of correspondence, and thereby arrive at certain general conclusions important for the general theory of scientific cognition ; and secondly, I want to use the characteristic of adaptation as the theoretical point of departure for a research on the historical-humanist cognition of arts, as this approach reveals some properties of the object that the art cognition is concerned with, namely, the process of development of the social artistic practice.

In other words, the last essay returns, by focusing on the problems discussed earlier, to the questions taken up in the first two essays: to methodological issues connected with historical cognition.

This review of the contents of the successive essays indicates that their topics are closely connected, not only in the sense expounded by in the title of this book. Naturally, they are all concerned with problems that belong either to the theory of historical cognition or to the general theory of scientific cognition. In the former case, the concern with cognition is double-layered : epistemological cognition is a form of historical cognition and therefore its results apply to its own practice as a form of historical cognition. But the connection between the essays is also of another nature. Each successive essay assumes as established the findings that have been made in previous papers. Consequently, the collection shall be read in the order in which it has been compiled.

Now, I must offer a commentary to a term frequently used above: theory of scientific cognition. To my mind it refers to every form of reflection on science, more or less rigorous, as long as it is philosophical in character, and, what is the most important, if it codifies and makes systematic the contents of the social methodological conciousness. The latter is the subjective-social regulator of research practice, or, more frequently, of a certain branch of that practice. Obviously, the question, how the research tasks of the theory of scientific cognition are formulated, depends on the philosophical orientation within which the question has been formulated. It is often claimed, as it seems, that the main research task is to answer the question: Which (scientific) cognition is valid ? It is also held, often simultaneously, that the answer should not (or even cannot) be based on the findings provided by particular science, and consequently the existing research on theories has a definite speculative-aprioristic character. Sometimes, e.g., in the case of epistemology of classical positivism, the problems of theory analysis are solved by reference to psychophysiological data. Sometimes, the case of hypothetism is the best example of that tendency, it is assumed that the task will be completed by reconstruction of norms and directives that are actually applied in research practice. In this case, it is also assumed that the results of the reconstruction are obligatory, i.e., the reconstructed norm and directives will not only be shown as in fact adhered to but also as legitimately binding. But no matter how individual epistemologists see the nature of their activities, its objective function is codification and systematization, and also (ideologically motivated) verbalization of specific elements of the social methodological consciousness at the given developmental stage of the research practice. As for Marxist theory of scientific cognition, it is taken for granted that it is based on historical materialism, i.e., that it is a history (in the sense opposite to the ordinary or positivist meaning of the term) of the social methodological consciousness — or to present the matter more thoroughly — a history of the development of the social research practice (as the science of the development of the social methodological practice must be the science of the development of the social research practice, or else it ceases to be based on historical materialism). I must additionally mention that I consider the term 'methodology of science' as equivalent to 'theory of scientific cognition', but will consistently use the latter name only, as normally the former is given a different meaning from what I assign to it, and besides it is used in a number of different senses. The upshot of all this is that what I call social methodological consciousness, especially in its normative part, can sometimes be verbalized philosophically (reconstructed) in the form of certain contents of that consciousness, e.g., as a logical theory of science.

In many cases, the papers included in this volume, or their parts, have been published before as articles in *Studia Filozo-*

ficzne or *Studia Metodologiczne*. In this edition, they appear in a thorougly examined, new form. Not only because I wanted to lend them an appearance of a compact and coherent whole, but also because I wanted to include the results of my recent investigations which I have been conducted after the original publications came out. These results have often been inspired by discussion with my colleagues engaged like myself in the study of methodology of historical research (i.e., research on developmental processes). I must say that the theses professed in this book are far from complete, they still need to be formulated more precisely or to be complemented by historical-scientific analyses. It is conceivable that many of them will have to be modified in the future, and some, perhaps, will be rejected. It is still an open question, I believe, what is the best form, i.e., the most consistent and the most convincing one, in which to present the fundamental assumption of these papers to the effect that epistemological cognition is a type of historical cognition.

NOTES

[1] J. Kmita, *Z metodologicznych problemów interpretacji humanistycznej* (*Methodological Problems in Humanistic Interpretation*), Warszawa, 1971.

[2] Strictly speaking, the position bearing the new appelation is not an old thing sold under a new name. Later on I will have more to say about this new position and some of its parallel variants. Then it will become clear that what is now called methodological anti-individualism was referred to in *Z metodologicznych problemów interpretacji humanistycznej* as structuralism or (methodological) holism.

[3] The distinction between idealizing humanist interpretation and concretized humanist interpretation has been introduced in my article 'Kilka uwag o idealizacji w badaniach logicznych nad językiem nauki' (Some Remarks on Idealization in Logical Studies on the Language of Science) (*Studia Semiotyczne*, Vol. 3, Wrocław, 1972). However, I have not emphasized the view that idealizing explanation is an explanation of the fact that the given activity was undertaken starting out, to put it briefly,

from the premises which express the point of view of social consciousness.

⁴ An additional question arises about the legitimate coupling of the beliefs of a given, concrete individual and one or another system of beliefs that belong to social consciousness. I will not take up this question here, although *Essays* contain several observations which can serve as premises to the formulation of an appropriate answer.

⁵ *Z metodologicznych problemów interpretacji humanistycznej*, cited above, p. 9.

⁶ The phrase 'penetrating into the realm of social consciousness', although not so elegant stylistically, is better than the term I used previously, viz., 'propagation'. The former has the advantage of highlighting the fact that a belief belongs to social consciousness, whereas the latter sugests only that the belief is widespread. The second fact is a good (perhaps even the most characteristic) symptom of the first one, but the two can be conflated only on the grounds of an individualistically oriented definition of social consciousness, which assumes that social consciousness is identical with the beliefs shared by all or the majority of members of a certain social group.

⁷ Here, and throughout this book, 'objective' means existing (or, in case of states of affairs, obtaining), irrespective of whether it has been registered by either individual or social consciousness, or not.

⁸ A secondary demand with regard to another demand of functional priority is such a potential objective effect of activities of a given kind which must be present in order to satisfy the latter demand.

⁹ This concept of social practical knowledge will be characterized more precisely in the forth part of these *Essays*. I will only mention here that the concept was designed by A. Pałubicka, in her article 'Nauka— doświadczenie społeczne—praktyka' (Practice—Social Practical Knowledge—Science), *Studia Metodologiczne*, 1976 No. 4. The author claims in that article that the beliefs which constitute the contents of social practical knowledge do not require prior articulation nor acceptance by concrete individuals. I will rely below on this presumption in my interpretation of the role of the intellectual activity of concrete individuals which contribute to the formation of social consciousness.

¹⁰ A Hauser, *The Philosophy of Art History*, New York, 1959, pp. 8–17.

¹¹ Subsequently, they must be respected by individuals who engage in their individual activities, or else their activities become ineffectual. But we must rigorously distinguish between the efficiency of individual activities (which materialize the values individually espoused) and the efficiency of social practice (which materializes the values constituted within the realm of social consciousness). I must mention here that I formerly believed (e.g., in *Z metodologicznych problemów inter-*

pretacji humanistycznej) that the efficiency of individual activities, as a symptomatic manifestation of the fact that a correlated belief belonged to social consciousness, was the proof that there was an objective demand for that belief. Now I think the assumption of such equivalence is too strong, and also inadequate as an attempt to reconstruct appropriate ideas of historical materialism.

[12] A more insightful analysis of the hierarchic functional structure (with respect to a given global property) has been given by K. Zamiara, *Metodologiczne znaczenie sporu o status poznawczy teorii* (*The Methodological Significance of the Controversy over the Cognitive Status of Theories*), Warszawa, 1973, pp. 183–230.

[13] To be more precise, one should speak in this and similar contexts of the necessary condition accompanied by specific circumstances of other, appropriate kind. As this Introduction does no more than present the basic intuitions, which are further developed in the essays that follow, I forego the intention of making all requisite reservations.

[14] In *Z metodologicznych problemów interpretacji humanistycznej* I made the assumption that the Marxist functional-genetic explanation does not require any reference to the concept of the functional structure, as it has been outlined above, and discussed more thoroughly in the book mentioned (cf. pp. 140–141 and 169). My main reason for this was the fact that the appearance of particular beliefs in the minds of particular individuals is not determined by the objective demand made by a diachronic functional structure as it is constituted by social practice conceived as a whole together with its concomitant conditions. This fact has been, and continues to be, neglected in the history of Marxist thought. Now I have rejected the above-mentioned assumption and my decision has just been manifested by saying that the functional component of the functional-genetic determination can be characterized in terms of the diachronic functional structure, which represents social practice together with its concomitant objective conditions. I have done so after a deliberation which has led me to the conclusion that a more careful discrimination between social consciousness and individual consciousness permits one to fully acknowledge this neglected fact and yet to perceive that social consciousness is functionally determined by objective demands. No incompatibility arises at this point. One thing is the functional connection between social practice together with its objective concomitants and social consciousness (as a certain set of its substructures), another thing is the connection between social consciousness and individual consciousness. In the view being expounded here, the latter connection is based on the assumption of rationality.

[15] It can be mentioned, addressing the remark to the philosophers who count the hypothetistic theory of scientific cognition among doctrines

of logical positivism, that their opinion is a flagrant manifestation of ahistorical approach. They lump together epistemological orientations which, admittedly, have contemporaneous representatives, working not infrequently in close proximity, but they neglect the fact that these authors express subjectively different developmental stages of the social research practice. What is more important, some of the philosophers I have in mind apply the term 'logical positivism' to epistemological reflection of every hue and colour as long as it is presented as a scientific endeavor (the telltale sign of this intention is, allegedly, the use of formal logic) undertaken in order to study the process of emergence of scientific knowledge. One dreads at the thought that by this definition of 'logical positivism' even the Marxist theory of scientific cognition might be called one of its branches.

ON TWO KINDS OF EXPLANATION

1.1. INTRODUCTORY REMARKS

Although scientific research practice has undergone historical changes and, moreover, as it displays parallel variant in the synchronic perspective, one can argue thut results of research practice always assume the form of a set of statements arranged more or less aptly into a deductive system. Elements of such a system are connected, explicitly or implicitly, by the relation of logical entailment. In the case of empirical sciences, some of these connections, if they meet additional conditions, are at the same time the connections of explanation. In spite of the fact that such connections frequently appear in the body of research results because investigators have undertaken a conscious effort to perform the explanatory operation, it is by no means necessary that such an operation precedes every instance of the occurrence of explanation. For that purpose, it is sufficient that a particular set of research results contains two units of knowledge (expressed in simple or compound statements) of which it can be said that they are connected by the relation of explanation as defined by the principles of explanation accepted in the community representing a given research discipline at a given time.

Certainly, these principles of explanation (relativized to the social consciousness connected with a field of research practice at a particular period of time), these methodological directives, which identify a correct manner in which an explanation can be produced, are usually different for different relativizations. Philosophical reflection on science has neglected this fact almost entirely. It tends to reconstruct these directives in an absolutist fashion (though does so implicitly, as a rule, because it also tends to be normative in its approach) as if explanation were always subject to the same directives in all fields of research practice and in all stages of its historical development. It is a different question that these reconstructions are usually so general that their adequacy extends quite far, and may appear valid for various sets of methodological explanatory directives. It must be admitted that these reconstructions often remain valid for different research fields, or for different stages of research in the same field, or in both respects.

There is no reason to believe, it seems to me, that these abstract (reconstructive) characterizations of the explanatory activities are invalid. They can be correct, and, in particular, the most general, or the 'framework' characterization which comprises only such elements that are common to all methodological directives about explanation can be legitimate. It is only incorrect to extrapolate a methodological 'model' of explanation defined for a particular field — and therefore possessing no more that a limited historical application — on all empirical disciplines throughout history. As I have just mentioned, philosophical reflection on science is usually characterized by ahistorical normativism of approach to the processes of explanation, and becomes fallacious if the proposed model of explanation is not general enough. It should also be borne in mind that the most general model, as we are presently going to see, has very limited contents.

A fairly popular model of explanation now found in philosophy of science has been constructed by logical positivism and is based on examples drawn mostly from physics. It is incongruent, however, with patterns of explanation found beyond that discipline, for instance in biology, and most conspicuously in the humanities.

The observations that follow are intended as a comparison between two kinds of explanation, one represented by the logical positivist model, the other being something new, not yet extensively studied. I will start by describing the latter, and then I will proceed to discuss some problems arising in connection with this novel form of explanation.

*   *   *

I will discuss the problems below using realistic language rather that a nominalistic one (I will be concerned with objects rather than descriptions). On this assumption, every explanation will speak about an *explanandum*, which is either a regularity of a singular fact, atomic or molecular. I will focus on these two kinds of cases, and additionally assume about the explanatory procedure that :

(1) it is a research procedure which encompasses an answer (usually incomplete) to the question why this particular fact of this particular regularity has occurred ;

(2) it provides an answer, or a description of an *explanans* (which itself is also an object on my realistic assumption), in statements that can be empirically verified (are 'synthetic' statements) and constitute, together wih certain logical presumptions, a logical antecedent for the statement describing the *explanandum*, i.e., the statement about the fact or the regularity to be explained.

As I said before, there is a methodological connection between (1) and (2). Whenever the question is asked: 'Why S ?', and the answer given to it is found acceptable in the discipline to which the question belongs, then the statement S is logically (although enthymematically) entailed by the answer.

It is also required as a rule that the answer (a description of the *explanans*) contain at least one law if the *explanandum* is a fact, and at least two laws if the *explanandum* is a regularity. I must emphasize that I do not need these requirements because I think that they are not indispensable for the 'framework' concept of explanation, which has been adopted here. Moreover I believe that it is easier to answer the question: What properties belong to statements that are accepted as laws (in a given group of disciplines at a given time)? — when one knows what instances of explanation are accepted as valid, than to answer the question: Which explanations are accepted as valid? — by reference to a properly reconstructed conception of the law. In other words, it is more operative to define law using the concept of explanation than to go in the opposite direction, as is usually done.[1] On the approach adopted here, laws can be generally defined as units of scientific knowledge which, while not descriptions of particular facts, are found (at least implicitly) in the explanatory answer as part of the description of the *explanans*. These remarks, made with the view to elucidation of the explanatory procedure, bear indirectly on the characterization of the concept of law. This should not be surprising because, as we have said, the two concepts, of law and of explanation, are closely connected in philosophy of science.

It may be objected to what I have said above that by imposing the 'framework' conditions on the activity of explanation we have tacitly assumed a certain concept of law if only because we have been making use of the concept of regularity as one form of *explanandum*. This objection would be justified if my concept of regularity were in fact based on the concept of law, which has not been the case. In this essay, I mean by regularity a state of affairs such that (1) whenever a particular fact obtains (atomic or molecular) of type A, then another fact obtains of type B; or (2) that the relative frequency of facts of type B in the class of events of type A is: (2a) equal to p, or (2b) is either greater or less than p, or (2c) is contained in the interval between p and q such that $0 \leqslant p < q \leqslant 1$. I assume at the same time that the characterization of the paricular states of affairs of type A does not contain a strict identification of their locale in space-time.

The regularity described under (1) is usually called *deterministic*, the regularity described under (2) is called *statistical*. I do not take my formulation of these regularities as sufficiently precise for general purposes, but it will do for my essay, especially as I want to show that the concept of regularity may be used without any preconditions about law.

1.2. TWO KINDS OF EXPLANATION — AN ANALYSIS OF EXAMPLES

The deductive model of explanation, usually applied in physics, is widespread in philosophy of science. It is often believed that the model is universal, and should be adopted in all empirical sciences and in the humanities. Sometimes it is enlarged by addition of a probabilistic model as its complement, and the two models together are said to account for all instances of explanation found in empirical sciences.

As it should be clear from the remarks I have made so far, the probabilistic model of explanation is groundless. It is never used in the ordinary research practice of empirical sciences, and, consequently, we never meet with an explanation based on this model.[2] The deductive model is correct in the sense that in many instances, especially in mathematical-natural sciences, it is used to provide an explanation. Methodological directives with which these instances of explanation are connected lend them properties required by the deductive model. Yet, I would like to make two reservations, not so much about the model itself, as about the method of its presentation.

First, I must submit that contrary to the current opinion the model does not revoke the 'spitting image' of the cases it purports to represent by providing their explanation. Hence it would be useful to distinguish between different variants of the model, typical for variaus historical stages of the development of sciences (practically, I am talking about mathematical-natural sciences only). Secondly, it is usually present-

ed in such a manner as to suggest that it can be applied
with equal ease in all academic disciplines, at all stages of
their development. In point of fact, however, the kind of explanation which prevails in the humanities and in biology when
it is engaged in the study of the processes of evolution has
nothing to do with deductive explanation. It is very symptomatic that these two disciplines are quoted whenever the
probabilistic model of explanation is offered as a complement
to the deductive one, which is allegedly not capable of covering
these two fields. But as I said, these attempts are out of touch
with real research practice in biology and the humanities.

The model that I will propose presently as a more adequate
complement of the deductive model is itself a deductive model.
Therefore, in order to avoid misunderstanding, I will call the
old one the model of *unequivocal explanation* and the new one
the model of *historical explanation*. My further remarks will
make it clear why I have decided on this terminology.

Although, as I have said, the model of unequivocal explanation is widespread and well known. I will present a simplified
example of explanation using this very model. It will be convenient to have one such example to refer to when I compare
the unequivocal model with the model of historical explanation.

Suppose that the *explanandum* is the concrete fact that the
gravitational interaction between Earth and Sun is inversely
proportional to the squared distance between their centres.
This fact (incidentally, it has been described imprecisely, as
I should have spoken of the mean distance) can be explained
on the grounds of classical mechanics in many ways, e.g., with
the use of Kepler's third law. For our purposes, it will be
most convenient to rely on the law of universal gravitation
and its immediate consequence, which stipulates that if individual physical objects x and y are spherical, then the gravitational interaction between x and y is inversely proportional to the squared distance between their centres. This consequence should have been expressed in a more complicated

manner, as the conditional in which the law of gravitation is expressed does not only require that in the antecedent the individual objects x and y be spherical physical objects (or material points), but also that they be the sole elements of a relatively isolated system consisting of two bodies. Thus through abstraction and idealization it presupposes that no other interaction with physical bodies needs to be taken into account. Although this aspect of idealization (in fact, idealization comes into the picture more than once in this formulation of the law) is very important from the methodological point of view, it may be neglected in this essay as the conclusions that I intend to draw are independent of it.

The individual terms 'Sun' and 'Earth' will be replaced for brevity with letters a and b, and the predicates 'x is a physical spherical object' and 'the gravitational interaction between x and y is inversely proportional to squared distance between the centres of x and y' will be replaced respectively with '$P(x)$' and '$Q(x,y)$'. The *explanandum* of our example may be described as '$Q(a,b)$', and the simplified consequence of the law of gravitation takes the form

$$\bigwedge_{x,y} [P(x) \, \& \, P(y) \, \to \, Q(x,y)].$$

When we attach the so-called initial conditions '$P(a)$' and '$P(b)$', which say that Sun and Earth are spherical objects, we obtain a complete description of the *explanans* which logically entails the *explanandum* :

(i)   (1) $\bigwedge_{x,y} [P(x) \, \& \, P(y) \, \to \, Q(x,y)]$

       (2) $\dfrac{P(a) \, \& \, P(b)}{Q(a,b)}$

The predicate abbreviated as '$Q(x,y)$' should be expressed in principle by an equation that describes an appropriate function. Yet the fact that the predicates of physics are almost always a description of functions does not touch upon the problems that we are here concerned with.

It should be noted that the explanation expressed as (i) pertains to a particular fact, viz., the fact that $Q(a,b)$. But we can easily construct an example of an explanation of a regularity. We just need to abstain from Sun and Earth and speak of two elements of any conceivable planetary system. Let '$R(x,z)$' be the abbreviation of the predicate 'x is an element of the planetary system z'. An explanation of the regularity will have the form

(ii)   (1) $$\bigwedge_{x,y} [P(x) \& P(y) \rightarrow Q(x,y)]$$

(2) $$\bigwedge_{x,y,z} [R(x,z) \& R(y,z) \rightarrow P(x) \& P(y)]$$
$$\overline{\bigwedge_{x,y,z} [R(x,z) \& R(y,z) \rightarrow Q(x,y)]}$$

We can obtain two formulas of maximum generality from (i) and (ii) if we get rid of their specific contents. The explanatory premises and the descriptions of the *explananda* may be much more complicated, the explanation of the regularity does not have to be presented in the form of syllogism, etc.

Let us move on to our second example of explanation. The *explanandum* will be the phenomenon called *Bates' mimicry*. It consists in '(...) imitation of a species called the model by another species or more if the former has some properties which defends it from predators. For instance, species which exude a repugnant smell, possess venom or are equipped with stings may be imitated in colour or shape by species which oftentimes belong to a distant systematic category, but have no natural means of defense. In such a situation, the specimens of the imitating species receive protection from predators due to their superficial resemblance to the model.' [3]

Bates' mimicry has not been characterized above as either an element of a deterministic or one of a probabilistic regularity. The sufficient conditions for its occurence have not been specified, and neither are we told (indeed it would be more difficult to venture a guess about this question), not even in rough approximation, what the probability of its occurrence in given circumstances amounts to. However, if the matter is simplified, one can place this phenomenon in the framework of a (deterministic) regularity. But we have to remember that jeopardized species can imitate its model only if a mutation of an appropriate kind has happened, and besides '(...) the form or forms that imitate the model must inhabit the same area as the model and must be less frequent than the model. If the last condition were not satisfied, birds would come across the specimens of the species without the offensive smell more often (species of butterflies are being discussed — J.K.), and these, as imitators, would be defenseless.' [4] Taking notice of the elements of the sufficient condition for the occurrence of Bates' mimicry we can reconstruct the following regularity: Whenever a species population P lives in an area occupied by the species population P', and P' has a natural protection from predators that P does not have, and P' is more populous than P, and within P a mutation occurs that makes its representatives look superficially like specimens of P' thanks to new properties $C_1, ..., C_n$, and the mutants are not defective in any other respect, then, after a certain period of time, the specimens with properties $C_1, ..., C_n$ will gain numerical preponderance within the species population P.

There is no doubt that from the biological point of view the list of components making up the sufficient condition for Bates' mimicry, described as a regularity (from now on referred to as M) must be substantially extended before it can count at all as an adequate description of facts. For the purposes of our analysis, however, the accuracy of M in its present formulation seems to be sufficient.

What is the *explanans* of the regularity M? M is intended as a manifestation of the 'operation' of the law of natural selection. A description of the 'mechanism of operation' is at the same time a description of the most important element of the *explanans*. Natural selection '(...) can be described as a process connected with mating which constitutes an indispensable attribute of the latter. (...) By the process of mutation, different alleles arising from the process of meiosis and insemination are accumulated in every population in different combinations. Different combinations of genes have different selective values, i.e., they are differentiated by natural selection. The direction of selection is determined by the entire set of conditions in the environment at a given time, constituted by the live and lifeless nature. (...) As a result of (...) natural selection changes occur in the genetic pool. Prevalence of individual genes is altered, and the genes which formerly have been eliminated may now propagate.'[5] 'Nature eliminates unconditionally only such specimens which die before maturity or remain unable to procreate. All other produce offspring, but such predominate which propagate more profusely and thereby tilt the balance of genetic endowment in their favour.'[6]

Here we have a new regularity which we may call N. It produces regularity M, and therefore N mainly explains M. N can be described as follows: Whenever within a species population P in natural conditions W_1, ..., W_n mutants occur with properties C_1, ..., C_n, which are not shared by the rest of the population, and those specimens which have properties C_1, ..., C_n are better adapted to conditions W_1, ..., W_n than those specimens which do not have them, then, after some time, the specimens with properties C_1, ..., C_n will gain numerical preponderance in population P.[7]

It will be noticed that if we can attach to this description of regularity N, which is an integral part of the theory of natural selection, the following description of a regularity, which we will call O: Whenever within the species population P, which occupies the same area as a more numerous population P' — which has properties ensuring protection against common predators — mutants occur which by virtue of their properties C_1, ..., C_n resemble by their external appearance specimens of population P' then these specimens will be better adapted to life in these conditions thanks to properties $C_1, ..., C_n$ than the remaining representatives of the population P, which do not possess these properties — then N explains regularity M, that is, N together with O entail a description of regularity M.

To better elucidate this situation, let us use abbreviations for the regularities N, O, M, and assume that the description of the regularity N will also be called the principle of natural selection N. The theory of natural selection contains also other principles as it could be seen, for instance, in the fragment quoted from *Ewolucjonizm* by S. Skowron, which explains the phenomenon of retention of some current properties of a given population (by pointing out that these properties ensure the best adaptation of the representatives of the studied population to the existing natural conditions).

Let '$\Phi(P)$' be an abbreviation of the phrase : 'In the species population P mutants occurred with properties $C_1, ..., C_n$; Φ is here a predicative variable of the second order that predicates on predicates of the first order, P is, naturally, a predicate of the first order. Both variables have a predetermined range of possible concrete substitutions (one could consider them to be predicates equipped with certain parametric variables, but my approach seems to me more convenient). Let '$\Psi(P)$' be an abbreviation of the phrase: 'Population P lives in natural conditions $W_1, ..., W_n$'. The remarks made about '$\Phi(P)$' hold for this new formula as well. I will only mention that among the conditions $W_1, ..., W_n$ we will have in particular the fact that in the area occupied by P there is another population, more numerous, endangered by the same predators and having certain additional properties — especially those that are now found in the mutants of the population P. Finally, let '$\alpha(\Phi, \Psi, P)$' and '$\beta(\Phi, \Psi, P)$' be abbreviations of the phrases: 'Specimens of the population P equipped by mutation with properties $C_1, ..., C_n$ are better adapted to conditions $W_1, ..., W_n$ in which lives the population P than other representatives of P which do not possess the properties $W_1, ..., W_n$', and 'When some time has

elapsed from the moment that mutants with properties C_1, ...,
C_n occurred and the conditions W_1, ..., W_n prevail, the specimens of P which possess properties C_1, ..., C_n will have gained numerical preponderance'. Expressions α and β are predicative constants of the third order. For the predicative constant of the second order that can be substituted for Ψ, we
will use symbol A; '$A(P)$' is to be read as 'the area occupied
by the species population P is also occupied by the species
population P' which is protected from predators common to
both populations, and P' is more numerous than P, and representatives of both populations have the same external properties C_1, ..., C_n'.

A shorthand explanation of the regularity M will look as
follows:

(iii)

$$(N) \bigwedge_P \bigwedge_{\Phi\Psi} [\Phi(P) \,\&\, \Psi(P) \,\&\, \alpha(\Phi,\Psi,P) \rightarrow \beta(\Phi,\Psi,P)]$$

$$(O) \bigwedge_P \bigwedge_\Phi [\Phi(P) \,\&\, A(P) \rightarrow \alpha(\Phi,A,P)$$

$$\overline{(M) \bigwedge_P \bigwedge_\Phi [\Phi(P) \,\&\, A(P) \rightarrow \beta(\Phi,A,P)]}$$

As we can see, explanation (iii) corresponds to explanation (ii).
In both cases, the *explanandum* is a certain regularity. When
considering a specific concrete case of Bates' mimicry, e.g.,
a population of butterflies *Basilaria archippus* imitating the colour patches on the wings of *Danais plexippus*,[8] we can obtain
an explanation of a concrete fact with the use of regularity M
in the following way:

(iv)

$$(M) \bigwedge_P \bigwedge_\Phi [\Phi(P) \,\&\, A(P) \rightarrow \beta(\Phi,A,P)$$

$$\frac{B(P_1) \,\&\, A(P_1)}{\beta(B,A,P_1)}$$

In this formula, the predicative constant B is one of possible
substitutions for the variable Φ and points to concrete colour
patches on the wings of butterflies *Basilaria archippus*, and P_1
is a predicate denoting a population of just those butterflies.

Let us note that the regularity M may be called upon to explain the fact that $\beta(B,A,P_1)$ only because this regularity can be explained in turn by the conception of natural selection (explanation (iii)). If the last regularity could be explained by a purely phenomenalist generalization (if such generalization can be found at all in contemporary biology), then, probably, the explanation of the form (iv) would not be considered satisfactory from the point of view of current methodological directives. Thus we can form the argument that scientific laws — which, as we assumed, must be eligible as satisfactory components of explanations — cannot be characterized in some stages of their development solely by phenomenalist generalization (if such are ever accepted as statements in sciences — to repeat once more the earlier caveat).

It must be emphasized that the fact that $\beta(B,A,P_1)$ holds can be immediately explained by the regularity N. In an abbreviated form this explanation, marked as (iv'), would contain two premises: a description of the regularity N and the statement:

(iv') $$B(P_1) \,\&\, A(P_1) \,\&\, \alpha(B,A,P_1)$$

1.3. UNEQUIVOCAL EXPLANATION VERSUS HISTORICAL EXPLANATION

When we compare statements (i), (ii) on the one hand, and (iii), (iv), (iv') on the other, we will not fail to notice a difference in the order of the predicates used (plus perhaps, but this circumstance is not important here, that they refer implicitly to different conditional tautologies). Statements (i) and (ii) contain only predicates of the first order. In (iii), (iv) and (iv') there are predicates and predicate variables of a higher order. This fact is not of primary importance as far as conclusions of my argument are concerned, but it could be a good starting point for an investigation of the correctness of methodological individualism with respect to the theory of evolution. If methodological individualism could provide a satisfactory account of the theory of evolution, then every statement concerning species populations, including most general statements, should be reducible for explanatory purposes to statements that can be formulated in the language of first order calculus of quantifiers — a language which says nothing about populations, but speaks about individual specimens. I find this possibility highly problematic as I believe in particular that the principle of natural selection, even in its fragmentary form which I gave it under the description of the regularity N, could not be reduced to statements containing only individual variables ranging over the elements of the set of individual organisms and first order predicates. I can corroborate this assumption with the following opinion of a biologist: 'The operation of natural selection (...) can be fully appreciated if instead of concentrating on individual specimens, we focus on their groups, i.e., populations. In fact, a population is a fundamental unit which undergoes evolutionary changes.' [9]

Now, the main difference that I wish to highlight between explanations (i) and (ii) on the one hand, and (iii), (iv) and (iv′) on the other, is that in the latter, but not in the former case, expressions of the lowest kth order ($k = 0, 1, 2, \ldots$; individual constants and variables are considering here as expressions of the zero order) in at least some instances are arguments of predicative variables (instead of constants) of the $(k + 1)$th order, and only these predicative variables are connected with predicative constants of the $(k + 2)$th order. Hence, in the explanations (i) and (ii), individual constants and variables (expressions of the zero order) are connected with predicates P, Q and R. In the explanations (iii), (iv), and (iv′), expressions of the first order, i.e., the predicate variable P and the predicate P_1, are connected with the predicates of the second order Φ and Ψ. These in turn are connected with predicate constants of the third order α, β. These formal differences represent an important difference of contents by contrasting two kinds of explanation against one another.

The descriptions of the regularities that we refer to in the explanations (i) and (ii) use unequivocal predicates, explicitly specified. In the explanations (iii), (iv) and (iv'), the predicates are not unambiguously specified; it has only been stipulated that they must satisfy condition a. This condition is directly stated in the explanations (iii) and (iv'), while it is no more than indirectly intimated in the explanation (iv), as in the last case the explanation of regularity M is reducible to the explanation of regularity N. Naturally, there can be many predicates that satisfy such condition which are not extensionally equivalent. Moreover, there is no effective procedure of their enumeration due not only to practical reasons but to theoretical reasons as well. The pairs consisting of recombinations of genes which generate specific changes in the organisms of a given population and the natural conditions in which this population lives — represented by predicative variables satisfying condition $a(\Phi, \Psi, P)$ (which says that specimens of population P which possess new properties are better adapted to their living conditions than the remaining specimens of same population — may have little in common for different species, and cannot be enumerated not only because the number of recombinations and different living conditions are impossible to exhaust, but also because they cannot be predicted for the future.

We may emphasize that specific cases :

$$\bigwedge_{P} \bigwedge_{\Phi\Psi} [\Phi(P) \,\&\, \Psi(P) \,\&\, a(\Phi,\Psi,P) \;\rightarrow\; \beta(\Phi,\Psi,P)]$$

of the general description of the regularity N:

$$\bigwedge_{P} [B(P) \,\&\, A(P) \,\&\, a(B,A,P) \;\rightarrow\; \beta(B,A,P)]$$

which have the form of descriptions of regularities contained in the explanations (i) and (ii), would be non-vacuously satisfied only with respect to one population at a given time. Besides, there is no reason to formulate such specific descriptions, which practically refer, in spite of the fact that variables are being used, only to a regularity obtaining in one case.

For reasons already stated, when we make use of regularity N in order to explain an evolutionary change, we must always apply concrete historical knowledge — in the broad sense of the term, meaning humanistic-and-natural knowledge — in the analysis to ascertain what properties belong to the population in question and what are the conditions in which it lives. These descriptions will provide concrete substitutions for the variables Φ, Ψ, which for the concrete circumstances satisfy condition a.

Let us call a *framework regularity* any regularity whose adequate description contains in the antecedent: (1) variables of the two lowest orders, k and $k + 1$, as an integral part of the description, and variables of the order k as arguments of the variables of the order $k + 1$, (2) constants of the order $k + 2$ which as arguments have the variables of the order $k + 1$, and the list of permissible concrete substitutions for the variables of the order $k + 1$ (substitutions which satisfy conditions stated in the constants of the order $k + 2$) cannot be exhausted for theoretical reasons.

We may assume furthermore that all other regularities that are studied in empirical sciences are of the same type as the regularities described in the explanations (i) and (ii), i.e., they contain expressions of the lowest order as arguments of predicative constants or contain an effectively produced alternative of such constants. These regularities will be called *determinate regularities*.

The concept will be useful in distinguishing between two elementary kinds of explanations : (1) *unequivocal explanation* whose *explanans* contains only determinate regularities — I must remind the reader that I am always using the objective style, and (2) *historical explanation* which contains at least one framework regularity. A historical explanation may refer to a framework regularity (as an *explanandum*), or may refer to a particular fact. In the latter case, we will speak of a particular historical explanation. Thus explanation (iii) is an explanation of a framework regularity, while explanations (iv) and (iv') are instances of particular historical explanation.

1.4. SOME CONTROVERSIES ABOUT THE METHODOLOGICAL NATURE OF HISTORY IN VIEW OF THE CONCEPT OF HISTORICAL EXPLANATION

Granting that the reconstruction of historical explanation, and especially of a particular historical explanation, is correct, we may quite easily understand the difficulties that many methodologists encountered when they assumed that in all empirical sciences only deductive explanation is used and refers to laws. The difficulty arises from the fact that by deductive explanation one normally understands an explanation which I have called unequivocal explanation, whereas explanations found in historical sciences, i.e., in the sciences which primarily rely on historical explanation, cannot be identified with this model without risking gross and absurd deformations.

Several responses to this difficulty have been made: (1) the deductive model of unequivocal explanation is complemented with a probabilistic model which plays the role of an all-purpose sack where to store every kind of explanation that does not fit the description of unequivocal explanation; (2) special concepts of inference are introduced, such as 'idiographic inference' of Donagan, which make it possible to retain the thesis of the deductive character of explanation, without, however, founding it on laws (in the narrow sense of the term); (3) explanations which differ from unequivocal explanations simply are not classified as explanations, and are either qualified as investigative procedures of a purely intuitionistic character which cannot be logically analysed (then instead of explanation we have a case of insight), or as attempts at explanation that must be rejected because they lack scientific rigour.

Arguments of the opponents of the thesis of the universal applicability of the deductive (unequivocal) model of explanation in historical sciences are usually more convincing as they refer to typical cases of explanatory procedures used by historians. Arguments of the proponents of the same thesis are usually derived from the peripheral areas of historical disciplines. In this case, unequivocal explanations are equated with common sense generalizations, which probably would not be accepted as laws by any sufficiently developed theoretical discipline. It must be doubted, in other words, whether such proposals are in fact explanations which satisfy methodological norms and directives socially accepted in the given discipline.

For instance, it is not difficult to show that in all cases when the action of a particular individual has been explained by pointing to the purpose that he/she wanted to achieve and to the knowledge that he/she possessed about how that purpose could be achieved, the connection cannot be expressed as a law, in the narrow sense of the term, a law which would connect the initial conditions of action with the action itself. From here it is only one step to a specific conception of understanding and to the conclusion that in humanist history '(...) the object to be discovered is not the mere event, but the thought expressed in it. To discover that thought is already to understand it.' [10] According to the position, however, which I defended in other of my publications, an 'understanding' explanation may only be defined as a deductive explanation which relies on laws. But laws are not conceived here in the narrow sense of the term, but as nomological formulae such as the assumption of rationality or efficiency of action.[11] These assumptions having the form of general conditionals must contain antecedents speaking about characteristic connections (predicative constants of the order $k + 2$) that should obtain between any knowledge (predicative variables of the order $k + 1$), any preference (predicative variables of the order $k + 1$), and any individual (variable of the order k) to explain why an individual acted as he/she did.

But even in the case when an explanation offered by a historian does not rely on the assumption of rationality or efficiency, and thus we are not faced with a case of humanist interpretation, some nomological formula rather than a law in the narrow sense of the term is to be reckoned with. For this reason, one must agree with the negative conclusions of the book by W. Dray, *Laws and Explanation in History*, where he criticizes the belief that historical explanation is an instance that falls under a deductive model of unequivocal explanation (covering law model). W. Dray shows conclusively that if we consider a typical historical explanation, we never directly extract from it a general and unequivocal law to be accepted by historians. For instance, a historian normally has no qualms about saying: 'Louis XIV died unpopular because of his pursuit of policies detrimental of French interests', but he certainly will not accept the general and unequivocal 'law' that 'Rulers who pursue policies detrimental to their subjects' interests become unpopular.' If pressed, the historian will agree that his original explanation is imprecise and too general, and he will consequently try to make it more specific by saying: 'Rulers who involve their countries in foreign wars, who persecute religious minorities, and who maintain parasitic courts, become unpopular.' But these specifications will still be considered too general, as one can easily find examples which are inconsistent with them. Finally, one reaches the point when no counterexamples can be cited to undermine the general statement, and it turns out that the general statement non-vacuously refers to only one case, that of Louis XIV.[12]

I believe that W. Dray's argument is perfectly sound. Any attempt to find an unequivocal law which underlies a historical explanation must end up with a general statement which is non-vacuously satisfied by only one instance. It must be so for

the same reasons which are responsible for the fact that particular unequivocal 'laws' reconstructed directly and solely
from the second premise in the explanations (iv) or (iv') about
evolutionary changes of the population *Basilaria archippus*
would refer non-vacuously to only one population probably.
I believe therefore that the proper reconstruction of the typical explanations used by historians must involve finding an
appropriate nomological formula. In the case of Louis XIV,
the nomological formula must characterize in its antecedent
those kinds of circumstances, primarily the form of social
relations, which made Louis XIV unpopular among his subjects. But even detailed enumeration of these circumstances
will not lead, via appropriate generalizations, to a reconstruction of a law in the narrow sense of the term.

It also seems that the premise which presupposes that the
only kind of laws are laws in the narrow sense of the term, and
the only kind of explanation is unequivocal explanation, is the
foundation of the idiographic conception, and especially of the
view that historical research is radically idiographic. This view
holds that historical research does not formulate laws, or even
that historical research has no use for explanation.

W. Dray, who shares these assumptions, does not, as a matter of fact, espouse radical idiographism, as he concedes that
historians make use of explanatory procedures. All the same,
he has adopted idiographism insofar as he claims that historians
have no use for laws in their explanations and do not formulate laws if only because they do not rely on them. It
needs to be pointed out that this view of the author of *Laws
and Explanation in History* is not based on a widespread belief
that historical events, unlike events studied by natural sciences,
are unique and never recurring, by virtue of their ontological
attributes. And yet 'Showing that there is no metaphysical barrier to bringing historical events under laws is not
the same as showing that the laws are in fact used, or that
they are in practice available, or that they must function in
the covering law way.' [13] In the research practice of historians,

'Historical events and conditions are often unique simply in the sense of being different from others with which it would be natural to group them under a classification term — and different in ways which interest historians when they come to give their explanations.' [14]

Similar theoretic premises, and namely concentration on laws in the narrow sense and on unequivocal explanation at the expense of nomological formulas and historical explanation, underlie K. Popper's arguments against 'historicism'. In *The Poverty of Historicism* Popper says of historicism that it assumes the existence of determinate regularities, to use my own terminology, which correspond to individual developmental changes in the history of human kind. Such regularities and laws (in the narrow sense) cannot exist because each would have to refer to an individual case. It is characterstic that those non-existent social laws about developmental changes are compared by Popper with non-existent, in his opinion, laws which control individual evolutionary changes in live nature. '(...) although there is no reason why the observation of one single instance should not invite us to formulate a universal law, nor why, if we are lucky, we should not even hit upon the truth, it is clear that any law, formulated in this or any other way, must be tested by new instances before it can be taken seriously by science. But we cannot hope to test a universal hypothesis nor to find a natural law acceptable to science if we are for ever confined to the observation of one unique process. (...) As applied to the history of human society (...) our argument has been formulated by H.A.L. Fisher in these words: "Men (...) have discerned in history a plot, a rhythm, a predetermined pattern (...). I can see only one emergency following upon another (...), only one great fact with respect to which, since it is unique, there can be no generalizations".' [15]

Popper's arguments would be convincing only if we spoke about laws in the narrow sense of the term. In this case, laws of developmental changes that would cover all cases of

changes which occurred in biological evolution on the one
hand, and in the history of human society on the other,
would have a range which comprises, as a rule, no more than
one case. The situation is entirely different if we take into
account laws which I called nomological formulae. When they
are admitted, the arguments of the author of *The Poverty of
Historicism*, intended, as it is widely known, against 'economic
historicism', i.e., against historical materialism,[16] fall apart.
The fact that specific historical explanations which rest on
Marxist theory of social development contain an *explanans*
which comprises a unique combination of historical events
does not rule out the possibility of formulating a law in the
form of a nomological formula which covers such a combination. In my opinion, the laws of historical materialism are
nomological formulae like, for instance, the principle N of
natural selection in the biological theory of evolution.[17] For
reasons that have already been discussed, explanatory application of these formulae to concrete situations presupposes a
historical analysis of the concomitant circumstances comprised
in the *explanans*.[18]

At least two nomological formulae are embedded, for instance, in the well known presupposition of historical materialism: 'Social existence determines social consciousness'.
We can see that nomological formulae were implied in that
formulation, and we can better understand how they should
be made explicit if we consider the following remark made by
F. Engels in his letter to Konrad Schmidt, concerning, first of
all, the problem of the development of philosophy, but pertaining also to the development of other areas of social consciousness: '(...) as a definite sphere in the division of labour,
the philosophy of every epoch presupposes certain definite
thought material handed down to it by its predecessors, from
which it takes its start. And that is why economically back-
ward countries can still play first fiddle in philosophy: France
in the eighteenth century as compared with England, on
whose philosophy the French based themselves, and later

Germany as compared with both. But in France as well as Germany philosophy and the general blossoming of literature at that time were the result of a rising economic development. I consider the ultimate supremacy of economic development established in these spheres too, but it comes to pass within the limitations imposed by the particular sphere itself: in philosophy, for instance, by the operation of economic influences (which again act only under political, etc., disguises) upon the existing philosophic material handed down by predecessors. Here economy creates nothing anew, but it determines the way in which the thought material found in existence is altered and further developed, and that too for the most part indirectly, for it is the political, legal and moral reflexes which exert the greatest direct influence on philosophy.' [19]

As we can see, there is a striking resemblance between what the contemporary biological theory of evolution calls a received pool of gene combinations in a given species population — which can be enriched by recombinations produced by mutation, and which is used in small proportion only due to selection by the conditions of the environment of such gene combinations which ensure that the specimens of the given species population will have the highest chance of adaptation — and what F. Engels called the heritage of the thought material of a given form of social consciousness. Keeping this analogy in mind and accepting a further consequence, namely, the analogy between phenotype ramifications of the genetic endowment of a species population which adapt it better or worse to natural conditions and man's social practical activity as responding more or less adequatly to the objective social demands in the configuration of the specific objective conditions (and further affected in the subjective aspect by social consciousness manifested in individuals as the expected result of the activity, i.e., as a purpose of and knowledge about the activity to be undertaken [20]) we can formulate two nomological formulae:

(1) Whenever within the class C of individuals characterized by their position in the social-economic structure emerges a new set of beliefs S which subjectively motivates a new practice, more effective in that structure (as a means of satisfying objective demands) than the practice subjectively motivated by beliefs S' socially accepted in the class C, then, after some time, the set of beliefs S will dominate in the class C in the said structure and will eliminate the beliefs S'.

The nomological formula (1) corresponds to the principle N of natural selection. The second nomological formula corresponds to another principle of natural selection which has not been discussed here so far.

(2) Whenever within the class C of individuals characterized by their position in the social-economic structure in the sphere of social consciousness dominates a set of beliefs S which subjectively motivates a practice that in that structure is more effective with respect to the objective demands than the marginally existing in C other bodies of beliefs, then, if the structure does not change, the body of beliefs S will dominate in class C.

These two nomological formulae are imperfect due to some simplifications. I will discuss two of them.

First, the social-economic structure (or the mode of production) is dynamic: if a certain practice becomes more popular, it must undergo a more or less radical change (similarly, an evolutionary change that characterizes a given species population alters the natural conditions of the population's habitat). Likewise, a prolonged viability of the practice changes quantitatively its structure. Now it must be pointed out that social consciousness never fits perfectly the conditions of social existence, but transforms itself in a process that never reaches its end. For that reason, individual consciousness of the representatives of a given class, if we leave infrequent exceptions apart, never reaches a state which fits in the theoretical framework of the social consciousness of that class.

Secondly, we can say in conformance with the just quoted excerpt from the Engels' letter that there can be different distances, so to speak, between various forms of social consciousness and related social-economic structures. Some forms of consciousness function practically in a direct manner (i.e., through practice they produce specific objective results in the structure); other forms of consciousness function more indirectly — through intermediate forms of consciousness closely connected to the economic underpinnings of the forms of consciousness (the consciousness which motivates productive activity affects legal and political consciousness, and this in turn influences ethical, religious, artistic, etc. consciousness). Thus, a practice which is motivated by forms of social consiousness lying at a greater distance from the economic background affects the forms of consciousness which lie closer to the structure, and thereby indirectly modifies the structure.

These two considerations should be kept in mind when historical changes of continuities are studied in the particular form of social consciousness of a specific period. This kind of explanation has been called by me a functional-genetic explanation [21] in order to emphasize the fact that such explanation focuses on the objective functioning of certain bodies of beliefs (through practice), and on the source of these beliefs (transmitted by tradition as thought material — to use a phrase of F. Engels). These two properties can be found also in the theory of evolution which explains changes and continuities in certain species populations ; the genetic aspect is connected with the gene pool, and the functional aspect concerns adaptive functions of certain properties which are transmitted through appropriate gene combinations in the species population. The biological concept of adaptation would have a social counterpart in the concept of efficiency of practice. As we can see, Engel's comparison of the importance of Marx' work for the science of social development to the importance of Darwin's work for the science of biological development has a deeper sense than is usually allowed.

NOTES

[1] This is a reverse procedure which relies on *a priori* characterization of the concept of law that makes no use of the concept of explanation. Then the concept of explanation is defined with the help of the concept of law. The procedure is not conveniently operational, and one reason for this is that nobody can naturally think of other indicators of the fact that a certain community of researchers has accepted a strictly general statement as a law than their readiness to use the law in explanation. If we make no use of this indicator, we will have to be guided by the observation that certain statements are as a matter of fact called 'laws'. This indicator is not very instructive if we take into account historical plurality and variability of meaning of the term 'law'. The concept of explanation is not open to similar objections. One can analyse particular cases of the activity of explanation (instances of answering the question 'Why S ?') and pay no attention to the manner in which the term 'explanation' is used, or even whether it is used at all.

We can note that the decision about the order of reconstruction of the concepts of explanation and law determines in particular certain types of arguments employed in philosophical analysis of science. The proponents of the order: first law, then explanation, will say that explanation of the fact that this bird is white by the fact that it is a swan and by the regularity to the effect that all swans are white is invalid. It is invalid even if we grant that all swans are white because the regularity stated in such terms is atheoretical, has a purely phenomenal character. From the point of view that I have proposed here, only the reverse argument can be accepted: a given statement cannot be considered law because the explanation connected with it cannot be accepted as valid. Naturally, in philosophical analysis of science one speaks more straightforward: this explanation is (is not) valid, this statement is (is not) a law. But if this normative style is to have a cognitive value, it may have it only as a certain, quite misleading manner of reconstruction of methodological norms and directives socially accepted in research practice.

[2] In this respect, I still hold the view presented in Z *metodologicznych problemów interpretacji humanistycznej* (*Methodological Problems in Humanistic Interpretation*), Warszawa, 1971, pp. 18, 19).

[3] S. Skowron, *Ewolucjonizm* (*Evolutionism*), Warszawa, 1967, p. 230.

[4] *Ibid.*, p. 230.

[5] *Ibid.*, p. 225.

[6] *Ibid.*, p. 226.

[7] The text quoted above seems to suggest that the phrase 'a better adapted variant of the species' means 'this variant of the species whose

members multiply in larger numbers than other representatives of the same species'. If this suggestion correctly expresses the actual point of view of contemporary theory of evolution, a rather complicated problem arises whether a more careful formulation of the principle of natural selection would not turn it into an analytic statement. I cannot investigate this question thoroughly in this essay because it is connected with a number of very controversial general issues. We would have to scrutinize the concept of analyticity (test the premises on which its standard version is based). We would have to inquire into a number of fundamental questions concerning the status of scientific laws and the relation of statements of mathematics to empirical statements. I will therefore only express my opinion that the principle of natural selection can probably be decomposed into two conditionals. The first will have the consequent which will be identical with the antecedent of the second one. It says that better adaptation is a sufficient condition of more profligate multiplication The second conditional combines profligate multiplication with numerical preponderance within the species. This second conditional may have a character that is commonly denoted by the concept of analyticity, as it is understood in methodological studies. In any case, I believe that K. R. Popper is wrong when he says in *Objective Knowledge* (Oxford, 1973, pp. 69, 70) that the concept of natural selection is a 'logical platitude'.

8 Cf. S. Skowron, cited above, p. 321.

9 *Ibid.*, p. 197.

10 R. G. Collingwood, *The Idea of History*, Oxford, 1946, p. 214.

11 Cf. e.g., J. Kmita, cited above and 'Kilka uwag o idealizacji w badaniach logicznych nad językami' (Some Remarks on Idealization in the Logical Studies on Language of Science), *Studia Semiotyczne*, Vol. 3.

12 Cf. W. Dray, *Laws and Explanation in History*, Oxford, 1957, pp. 31–34.

13 *Ibid.*, p. 46.

14 *Ibid.*, p. 47.

15 K. R. Popper, *The Poverty of Historicism*, London, 1966, p. 109.

16 Cf. K. R. Popper, *The Open Society and Its Enemies*, London, 1947, Vol. 2, p. 94.

17 The analogy between Marx' characterization of the regularities of social development and the regularities of biological development was emphasized in a review of *Capital* published in *Vestnik Evropy*, Petersburg. Karl Marx included this review in the annex to the second edition of *Capital* as an adequate presentation of his main theoretical-methodological ideas.

[18] Some of the intuitions that I have associated with the concept of nomological formula were expressed earlier with the use of the concept of essential historical generalization. Cf. 'Sens i użyteczność dziedzictwa naukowego klasyków marksizmu' (The Sense and Usefulness of the Scientific Heritage of the Founders of Marxism), *Nurt*, 1973, No. 2.

[19] F. Engels, 'To K. Schmidt', in: K. Marks and F. Engels, *Selected Works in Three Volumes*, Moscow, 1977, Vol. 3, p. 494.

[20] '(...) At the end of every labour-process, we get a result that already existed in the imagination of the labour at its commencement. He not only effects a change of form in the material on which he works, but he also realizes a purpose of his own that gives the law to his modus operandi, and to which he must subordinate his will.' K. Marx, *Capital*, Moscow, 1954, Vol. 1, p. 178.

[21] Cf. J. Kmita, 'Uwagi o holiźmie marksowskim jako koncepcji metodologicznej' (Remarks on Marx' Holism as a Methodological Conception), in *Założenia metodologiczne 'Kapitału' K. Marksa (Methodological Premises of 'Capital' by K. Marx)*, Warszawa, 1970.

ESSAY 2

ON THE CONCEPTS OF HISTORICAL POSSIBILITY AND NECESSITY

2.1. INTRODUCTORY REMARKS

According to W. Dray, whom I quoted in the previous essay, '(...) there is an important distinction to be drawn between explaining why a thing happened and answering a certain kind of "how" question about it. In the latter case (...), the historian need not show that what is to be explained happened necessarily in the light of the particular events and conditions mentioned in the explanation, and, a fortiori, need not show that it happened necessarily in the light of some covering law or laws. For the demand for explanation is, in some contexts, satisfactorily met if what happened is merely shown to have been possible (...).' 'In explaining why something happened (...), we rebut a presumption that it need not have happened, by showing that, in the light of certain considerations (perhaps laws as well as facts), it had to happen. But in explaining how something could have happened, by showing that, in the light of certain further facts, there is after all no good reason for supposing that it could not have happened.' [1]

In the tradition of Marxist philosophical thought, the two concepts, of possibility and necessity, are not opposed to one another, but to still further concepts, and namely, what is possible is contrasted with what is actual, and what is necessary is contrasted with what is contingent. There is no serious reason, I believe, to accept unconditionally only these two pairs of opposites; the purpose of investigation should decide the issue. If one intends to analyse, say, the most general characteristics of the mechanism which turns into reality what has initially been no more than possible, and as such, only one of the alternatives competing for being made actual, then the best framework for the analysis is the opposition: possible vs. actual. Speaking parenthetically, I do not consider elements of this opposition mutually exclusive, as neither does W. Dray. What is possible encompasses also such states of affairs which have already been made actual and therefore is already part of the real world. However, we may be interested in the distinction between actual states of affairs (facts) which obtained of necessity, i.e., have been produced by certain regularities and concomitant circumstances, and actual states of affairs (facts) which have not come around out of necessity but by chance. Now more appropriate will be the opposition: necessary vs. contingent. I may add parenthetically that I take this opposition to be relative to a specific group of regularities. When we speak of a state of affairs s as contingent with respect to certain regularities, we are not saying that it is absolutely contingent, we are not saying that even when we know of no regularity which makes it necessary.

The subject of a philosophical reflection may also be a process in which what initially has only been possible turns out to be necessarily real, appears as a necessary fact. In this situation, the opposition: possible vs. necessary is of primary importance.

As the foregoing remarks have made it clear, I have assumed certain relationships between the concepts discussed. The scope of the concept of actuality is comprised in the scope of the concept of possibility. Additionally, the scope of the concept of actuality is equal to the set-theoretical union of the scopes of the following concepts: that which is contingent and that which is necessary, if the two are constructed as relative to one (the same) group of regularities. Elements of the scope of all these concepts are states of affairs, and some of them are actual or factual states of affairs, or simply facts.

This essay has two fundamental tenets, the first of which is that the two procedures identified by W. Dray in historical research are connected in the following way: the answer to the question (i): how was it possible that a specific, temporarily localized state of affairs, i.e., a single phenomenon corresponding to an atomic or a molecular statement with a suitable temporal coordinate could occur, or a series of such phenomena could occur in successive time periods? — is in certain conditions one of the fundamental premises of the answer to the question (ii): why that state of affairs had to occur, or why as a matter of fact did it occur? The answer to (i) can be no more than one of the premises for the answer to (ii), as I said, because it has to be complemented by further informations relevant to question (ii), as it has to be an answer which entails a statement to the effect that the explained state of affairs actually occurred. As I have pointed out before, I take it for granted that a statement to the effect that a certain state of affairs occurred necessarily constitutes a presupposition for the statement that the sentence which states the occurrence of this state of affairs is logically entailed by sentences with the following properties: some of them describe certain regularities; all of them constitute an answer to the question, why the incriminated state of affairs did occur. For various reasons, it may be impossible to meet the last condition, and therefore sometimes we are unable to answer the question: 'Why S?', even though we may, conceivably, have an answer to the question: 'How was it possible that S?' Oftentimes we cannot explain the state of affairs consisting in the occurrence S, even though we can answer the question how it was possible that S was the case. Inasmuch as this is true, we may agree with W. Dray that the procedure of finding an answer to the question: 'How was it possible that S?' is different from the procedure of finding an answer to the question: 'Why S?' Unlike W. Dray, however, I only call the second procedure explanation.

To make this thesis more precise, I will use two explications.

E_1. State of affairs s constituting a single event or a series of events in the previously explicated senses of these terms is *historically possible* from the point of view of knowledge K if knowledge K logically entails, when framework regularities (in the sense made precise in the previous essay) are taken into account, a non-tautologic alternative built of at least two components which are made up of sentences declaring correspondingly: (1) the state of affairs s, and (2) other states of affairs, $s_1, ..., s_n$, co-temporal with s and pairwise exclusive from the point of view of K. Moreover, the alternative has the property that if we eliminate one of the component, we obtain an alternative (in a special case, a one-component alternative) which is not logically entailed by K. I will say additionally that each of the states of affairs: $s, s_1, ..., s_n$, may themselves be alternatives.

On the grounds of this explication, it is clear that from the point of view of K the answer to the question: 'How was it possible that s?' must enumerate all relevant circumstances and framework regularities ascertainable by K in the light of which the occurrence of the state of affairs s was not only possible (if it were ruled out by K, it would not have been included among the components of the alternative defined in E_1), but was complementary to other states of affairs $s_1, ..., s_n$, which could not be *a priori* excluded. The last consideration is often neglected when the possibility of the occurrence of an event or a series of events is being characterized. It is evident, however, that if we do not want, or cannot, prove the necessity of an occurrence or a process, we presume that another occurrence may have happened, excluded by and complementary (in a non-logical sense of the term) to the occurrence we have in mind.

E_2. The state of affairs s is historically possible from the point of view of knowledge K and recognizable as a state of affairs *historically necessary* on the grounds of knowledge K' if K' comprises K, and the statements in K' which are not at the same time in K specify circumstances which are not-antecedent to the circumstances registered by K plus, optionally, certain framework regularities. These residual statements jointly rule out states of affairs $s_1, ..., s_n$, which are possible on the grounds of K as well as s.

It can be stated from the point of view of the possessor of knowledge K subsequently transformed into knowledge K' that as a result of the occurrence of some additional circumstances the historical possibility s has changed into historical necessity.

Thus, the first tenet of this essay is that historical explanation of a specific state of affairs s which is either a single event or a series of them consists, at least on certain occasions, in the showing how the state of affairs s has changed from a single possible event into a necessary event. To illustrate this thesis, I will outline presently a number of examples of such explanations from historical researches. In Section 2.2, I will work out the second main tenet of this essay, viz., that Marxist explanation of social developmental processes proceeds by demonstrating how historical possibilities change into historical necessities. The terms of historical necessity and possibility are used in a specifically Marxist sense by imposing on these general concepts certain characterstic conditions.

2.2. TRANSFORMATION OF A HISTORICAL POSSIBILITY INTO A HISTORICAL NECESSITY — EXAMPLES

When I speak of historical research, I use the term in a wider meaning than is usually accepted. I give the name of historical research to every scientific procedure which is either a procedure of historical explanation or leads to such explanation. Historical explanation is understood in a sense that has been established in the previous essay : it is an explanation which contains in the *explanans* at least one framework regularity —· or to use the metalanguage style — which contains at least one law in the broad sense of the term as a nomological formula. I also wish to mention that if the subject of an explanation which comprises at least one nomological formula in the *explanans* is a state of affairs which itself is not a regularity, the predicative variables which belong to the formula must be substituted with constants which satisfy the conditions introduced by the predicative constants of the higher order of the appropriate nomological formula. Explanations of this kind, which I call particular historical explanations, must be based on an analysis of concrete circumstances which accompany the event to be explained, as we are not effectively informed in advance by the nomological formula in question what substitutions should be made for predicative variables. If we want, for instance, to explain, making use of particular historical explanation, a state of affairs which consists in the fact that in a concrete species population the number of mutants with certain properties has increased, we must determine those properties and conditions in which this species population subsists. Neither information is given in advance. If we find out subsequently that the mutants are better adapted to those conditions than other specimens of same species population, we may infer, relying on an appropriate nomological formula, that in the studied population mutants will tend to prevail.

This example brings to mind a fitting illustration of the first tenet of this essay, viz., that historical research consists very often in demonstrating how a certain state of affairs has changed from a historical possibility into a historical necessity.

The example shows that as long as we consider gene combinations in a given species population and gene recombination in the mutants, starting from the nomological formula of natural selection, we may only identify a number of alternative prevailing gene combinations complemented by the new mutant recombination. The gamut of these possibilities is reduced by the restriction that we are interested in a population which lives in a specific natural environment and by the assumption that the environment will not change. As a rule, it is clear on these preconditions that the currently prevailing variant of the species will retain its status, or, on the contrary, that it will be eliminated by specimens which represent a new, mutant recombination of genes. In this way, we can establish that one of the possibilities discovered by a concrete historical analysis will change, with the help of an appropriate framework regularity, into a necessity.

The environment may also change, of course, and the new conditions will favour a new possibility, which then will have a chance to turn into a necessity — a new variant of the species will prove better adapted to the new conditions.

Let us turn to a new example of transformation of a historical possibility into a necessity, when the former is characterized by each of the permissible and historically relevant substitutions for appropriate predicative variables in the nomological formula. This example belongs to the humanities.

Let us consider a set of possibilities representing activities one of which only can and will be undertaken by an individual. The individual views these activities through his/her subjective knowledge as states of affairs that are mutually exclusive and jointly complementary. These possibilities are conducive to a set of results which are expected outcomes as indicated by the subjective knowledge. Each possibility produces only one outcome. If we know nothing more about the individual in question — he/she may be a poet trying to find a suitable form of artistic expression or a military commander who must choose an efficient tactical move, etc. — then we must think in terms of alternative possibilities. The selection of one of them, the choice which turns one of them into a fact, depends on additional circumstances. They are determined by (i) the knowledge about preferential ordering of the expected outcomes characteristic for the choosing individual at the time of his/her decision-making, and

(ii) a nomological formula, which in this case is the assumption of rationality stipulating that in all circumstances such activity is selected which, from the point of view of the subjective knowledge of the agent, leads to the most preferred outcome.

In this case, one of the possibilities transforms itself into a historical necessity in accordance with the framework regularity corresponding to the assumption of rationality and the actual preferential ordering of the outcomes that can be produced by the activities to be undertaken. The explanatory procedure which is based on these premises and is employed in order to answer the question, why this particular activity has been selected from among those which were available, is called by me humanistic interpretation.[2]

The third example leads us to a characterization of Marxist concepts of historical possibility and necessity. These concepts are stronger, as I said before, than their ordinary counterparts, as they have more contents — they are ordinary concepts additionally strengthened by special conditions.

We will see it more clearly after we have discussed the third example. I will start by presenting it in a general outline, then I will discuss it in detail.

Let us consider an alternative of possibilities which represent different developmental variants of a social practice of a given type. I am afraid I cannot stop now to explain what I mean by a 'type of social practice'. It will be enough to say that it is an aggregate of practical activities performed universally and conducive to a common, widespread, objective effect of a certain kind. The effect is objective in the sense that it has been achieved irrespective of whether it was an intended effect (a consciously adopted purpose) in the minds of its instigators, or even irrespective of whether they had given it a thought. As we can see, when one endeavors a characterization of a given social practice in order to identify its type, one must take into account objective conditions which surround this practice as they determine the widespread, objective effect.

This global, objective effect has a crucial role in the determining of the set of alternative possibilities of further development of social practice of a given type. It determines the state of that practice in the society in question at a particular time and the manner in which the number of the initial possibilities will be eventually reduced. The connection between the direction of development of a social practice of a given type should be construed in the light of two opinions that follow, the first coming from Engels, the second from Marx.

'(...) we simply cannot get away from fact that everything that sets men acting must find its way through their brains. (...) The influences of the external world upon man express themselves in his brain, are reflected therein as feelings, thoughts, impulses, volitions in short, as "ideal tendencies", and in this form become "ideal powers".' [3] It is important for us that all practical activity has a subjective conscious motivation, which, for my part, I would tend to incorporate in the framework of the antecedent of the assumption of rationality as it comprises knowledge of alternative available activities, of their results, a specific preferential ordering of the outcomes, of one of them perceived as the sense of the next activity to be taken up, etc. This complex system of subjective conscious motivations is naturally correlated with every type of social practice. I am not talking now about subjective motivation of an individual, I am talking about the subjective context of the activities that are universally performed in a given social group. This context is a fragment of social consciousness connected with the group in question.

Hence, the second improtant factor — apart from the widespread, objective effect — which characterizes the type of social practice is its subjective social context. I will note in passing that the term 'subjective' is used by me in the sense 'connected with consciousness', and not in the sense 'contained in an individual consciousness'. The latter sense is included in the meaning that I impose to this term, but in my usage, what is subjective can be subjective either individually or socially.

Let me point out that the subjective context of social practice of a given type cannot be interpreted as a set of those beliefs which are displayed by individuals making up a group we are interested in. Admittedly, we might register these beliefs and inductively aggregate them to claim that the result constitutes the contents of a socially subjective context. This procedure, however, would only be possible on the grounds of methodological individualism, which is incompatible with the tenets of historical materialism.

The subjective social context of a practice of type T contains similar components as the subjective individual context: i.e., preferential ordering, knowledge which determines implementation of this ordering, etc. In theory, the subjective social context may remain in two kinds of relationship to the global, objective effect of the practice of type T: (i) the purpose (the value most preferred) cannot be accomplished even though some activities are subordinated to it — namely those which were selected subjectively from the point of view of social consciousness, and which by virtue of being universally performed are already part of social practice of type T; in other words, the effect rules out the achievement of the contemplated purpose; (ii) same purpose is tolerated, so to speak, by the objective, global effect of the practice of type T, which means that the subjective social context of the practice of type T poses an objective within the reach of the practice embedded in its context and the current objective conditions; in other words, the purpose can be achieved, and so its context will be called from now on an effective context.

As a matter of fact, situation (i) in principle cannot arise and consequently cannot last. In Marx's words: '(...) mankind always sets itself only such tasks as it can solve; since, looking at the matter more closely, it will always be found that the task itself arises only when the material conditions for its solution already exist or are at least in the process of formation.' [4]

This quotation from Marx is in my opinion a direct premise of the tenet about the determination of social consciousness by objective preconditions of social existence. Social consciousness contains only such beliefs that point to purposes and methods of their achievement lying entirely within the reach of, i.e., are permitted by objective conditions. This tenet, as I have already pointed out, should be construed as an abbreviation of a correlated nomological formula.

Thus, taking into consideration the state of social practice of type T in a given society at a given time, we can distinguish on the strength of the nomological formula about determination of social consciousness by objective conditions only such historically registered local variants $T_1, ..., T_n$ which as alternative possibilities of further development are accompanied by subjective social context that are effective.

These possible variants are not equivalent to one another in the long run, however. After a presumptive or actual proliferation of the variant T_i of the practice of type T, when the objective, global effect of practice T_i quantitatively grows, a qualitative change suddenly occurs: the subjective social context of the practice ceases to be effective, and thus its further continuation ceases to be feasible. This shows how a possibility is transformed into a necessity. Necessity is a former possibility which was left out as the only one possessing a developmental perspective not yet blocked and stopped.

We will study this process more carefully in the next section, but before we do, I would like to discuss briefly an example that illustrates the foregoing suggestions.

In the simplest case, there are only two possibilities of a further development of a social practice of type T: (i) continuation of the practice in its current variant T', (ii) replacement of that variant by a new xariant T'' which begins to emerge spontaneously at a given historical period, and for some time, the two variants, T' and T'', remain feasible. This means that their subjective social contexts are equally effective. As the objective, global effect of the variant T' grows quantitatively, however, its subjective social context loses its effectiveness — and the variant ceases to be feasible. Now, the possibility T'' becomes a historical necessity. It is theoretically foreseeable that the opposite possibility may have won the day. But development is possible only if the described course of events has taken place. This was also the only alternative that seriously commanded the attention of the first Marxist authors.

I may add that the blocking of a further continuation of a specific variant of a given social practice comes from dialectic contradictions within its bosom. I will have more to say about it later.

A concrete example of such developments can be found in the practice of exchange. Its earliest historical variant is exchange connected with the simple or, in Marx's terminology, single form of value.[5] Henceforth, I will call this variant simple exchange. It is epitomized in the formula: 'amount x of commodity A is equal to amount (value) y of commodity B'. The second variant, historically parallel and/or posterior, is exchange connected with the existence of, in Marx's terminology, the general form of equivalent value. In its final stage it is manifested as money. I will call this variant exchange with a general equivalent.[6] Now, prolonged continuation of simple exchange ceased to be feasible after some time and exchange with a general equivalent became a historical necessity. 'The gradual extension of barter, the growing number of exchange transactions, and the increasing variety of commodities bartered lead (...) to the further development of the commodity as exchange-value, stimulate the formation of money and consequently have a disintegrating effect on direct barter.'[7]

I will say below what dialectical contradiction, stemming

from the quantitative expansion of the objective global effect, blocked further continuation of 'direct barter', or simple exchange of commodities.

2.3. MARXIST CONCEPTS OF HISTORICAL POSSIBILITY AND NECESSITY

I would like to present in more detail my interpretation of the Marxist concept of social practice, then I will discuss dialectical contradictions implicit in any specific type of social practice, and finally wrap up the topic with a review of the concepts of historical possibility and necessity. I will illustrate my explications by returning to the example already offered for consideration.

As I pointed out, the concept of social practice of a given type is determined by two historically changing entities: the objective global effect and the subjective social context. We will focus on them now.

Activities which are repeated in the framework of a practice of type T usually have a compound character, i.e., they consist of more elementary components. An analysis of these activities and the relations established between their components will yield different results depending on whether we approach the subject from the objective or from the subjective point of view. If we perceive a compound activity from the objective point of view, i.e., when we see it as an *objective activity*, and leave unanswered all questions about the purposes and knowledge of the agent who performs the activity, then we are solely interested in causal relations. In this perspective, we (i) assign an autcome that is relevant to us to a given objective activity, and (ii) single out such components of that activity that are relevant with respect to the outcome in question.

Let there be an activity a which together with current objective conditions $c_1, ..., c_n$ constitutes the sufficient condition for the occurrence of the state of affairs b, and both a and $c_1, ..., c_n$ are not posterior to b and are essential components of the sufficient condition for the occurrence of b. The same can be expressed by saying that there is a causal relation between a and b, and that b is an effect of a. Certainly, both names representing events should be understood as relativized to conditions $c_1, ..., c_n$. One may, possibly, distinguish within a several components $a_1, ..., a_m$ which both, jointly and separately, enter into a causal connection with the resulting state of affairs (on account of conditions $c_1, ..., c_n$) when a enters into a causal relation with b (on account of conditions $c_1, ..., c_n$). I will call the components $a_1, ..., a_m$ direct causal components of activity a on account of the effect b and conditions $c_1, ..., c_n$. A similar operation can be performed in some cases on every component activity $a_1, ..., a_m$. We obtain thereby direct causal components of $a_1, ..., a_m$ (on account of conditions $c_1, ..., c_n$), and simultaneously indirect causal components of a. By iteration we may eventually single out the most elementary components of activity a.

Thus, a given compound activity a conceived of as an objective activity d_a (on account of effect b and conditions $c_1, ..., c_n$) is a directed graph whose nodes (a set of points) are instances of activity a, which produce effect b, and usually also instances of direct and indirect causal components of activity a (on account of effect b and conditions $c_1, ..., c_n$). The relation represented by the graph is conveyed by the predicate: 'x is a direct causal component of y on account of effect b and conditions $c_1, ..., c_n$'. In the extreme, simplest case, an objective activity is a directed graph described on a one-element set of points.

On the other hand, the same activity a may be presented as a humanist structure. It is then relativized to a certain sense, or purpose, s, and certain subjective knowledge K about the relation of causal connection which, according to K, obtains between activity a and purpose s, and about the relation of causal connection between causal components of a. Since the effect of the activity a, or the effect of its components, when viewed from the point of view of knowledge K, is a purpose

that should be accomplished, the relativized subjective relation of the causal connection will be called the relation of
causal subordination and the causal components of activity a
will be called direct or indirect instrumental components of
activity a.

Now, a given compound activity a conceived of as a humanist structure h_a on account of sense s and knowledge K is
also a directed graph whose nodes are activity a instrumentally subordinated to sense s, and (usually) direct and
indirect instrumental components of activity a on account
of sense s and knowledge K. The relation represented by
the graph is conveyed by the predicate: 'x is a direct
instrumental component of y on account of sense s and
knowledge K'. In the extreme, simplest case, humanist structure is a directed graph described on a one-element set of
points.

Practical activity will be understood as follows. A compound
activity a is practical activity only if it is an objective activity
d_a on account of effect b and conditions $c_1, ..., c_n$. If additionally it has a humanist structure h_a on account of sense s
and knowledge K, I will call it an effective practical activity.
If, however, activity a is an objective activity which fails to
bring about s on account of knowledge K, I will call it ineffective practical activity. A compound activity a which fails to
accomplish s on account of knowledge K may be viewed as a
special case of the so-called trivial directed graph in one-
element set of points which in this case represents nothing
but activity a. Such activity a will be called humanist quasi-
structure on account of sense s and knowledge K. A humanist
quasi-structure differs from a humanist structure based on
a one-element set insofar as the former is based on a point to
which no sense is instrumentally subordinated.

A *practical activity* p_a is therefore an ordered pair consisting of an objective activity d_a and humanist structure or
quasi-structure h_a. On the assumption that sense s and knowledge K have been unambiguously assigned to the practical

activity as its characteristic properties, the concept of practical activity may be defined relative to the parameters of the objective activity.

Let us note that in the extreme case of full adequacy of knowledge K and identity of sense s and effect b, objective activity d_a and humanist structure h_a are isomorphic. This isomorphism is established by the relation of identity.

I will give the name of the subjective context of practical activity p_a to the whole consisting of knowledge K, preferential ordering which co-determines sense s, and the relation of instrumental subordination of humanist structure (quasi-structure) h_a which is part of p_a. This context may be a part of social consciousness, or may be an individual case of an appropriate fragment of social consciousness. If so, I will call the correlated activity a *social practical activity*.

Two social practical activities which are relative to effects b and b' and to conditions $c_1, ..., c_n$ and $c_1', ..., c_n'$, respectively, are objectively homogeneous if they are isomorphic as objective activities in such a way that (i) corresponding elements of isomorphism of the two activities are of the same type, and (ii) corresponding effects b and b', and conditions $c_1, ..., c_n$ and $c_1', ..., c_n'$ are of the same type.

We will finish these terminological preliminaries with a remark about correct understanding of the fact that atomic or molecular states of affairs, i.e., components of objective activities, effects and conditions, belong to the same type. They do, if they enter into the same relations or possess the same properties, a fact to be witnessed by their description — assuming that a general order of describing states of affairs has been established. For instance, the two states of affairs: the fact that $Q(a)$ and the fact that $Q(b)$, are of the same type. It must be noted that the subjective context of social practical activity has its own type as well. This type, however, can be reduced neither to the sense, which is a specific, single state of affairs, nor to knowledge K, which is a compound consisting of concrete beliefs. The relativization in question can be found in many individual activities.

Let us assume that there are two sets of practical social activities which are effective and objectively homogeneous on account of, respectively, effects of type B and B' and conditions of type $C_1, ..., C_n$ and $C_1', ..., C_n'$. To each set corresponds at least one subjective social context, one type of humanist structure. Now, if states of affairs of type P are components of the effects of type B, and without them the correlated activities would be ineffective, we can say that these activities are an answer to the *objective demand* — they fulfil the demand of type P. If the same could be said about the activities belonging to the second set, then our conclusion can be extended and holds for the two sets. Objective demand is therefore connected with a given set of practical social activities and consists of those common components of the purposes of these activities which must be achieved or else the activities themselves are rendered ineffective. Two cases must now be distinguished. The activities belonging to the set in question may be actually performed, and then they satisfy the demand, or they may not be performed even though there are objective conditions which guarantee that they would have satisfied the demand if they had been performed. I will say that an objective demand exists if either of these cases is true.

My remarks should make it clear that all practical social activity is an answer to an objective demand, and the senses of the corresponding humanist structures are values which subjectively represent this demand.

The proposed manner of defining the concept of objective demand, connected, as I will presently show, with a given type of social practice, takes advantage of the circumstance that the salient consequence of leaving a demand unsatisfied is ineffectiveness of those activities in the first place which are part of the relevant type of social practice. It is possible to characterize the concept of objective demand in a theoretically more fundamental manner, as I did briefly in the Introduction. The entire social practice, together with concomitant objective conditions, should be seen as a diachronic, hierarchical, functional structure, whose global developmental property is the ability to reproduce old objective conditions and create new ones. We can say now that objective demands are (potential) collective effects of different types of practice, and that without these effects the global developmental property of the structure would eventually cease to exist. It would cease to exist eventually because particular types of social practice may be very remotely connected with the global developmental property. Many intermediary types of practice can be found between the global property and the objective demands of the social practice of a given type. If their demands are ordered $P_1, ..., P_n$, then demand P_1 will be directly subordinated to, and serve a function for, the global property; P_2 will meet the demand created by P_1, and so on, all the way until P_n is reached. The demand of the type of social practice that we focused on is directly subordinated to P_n.

It is obvious that before any other demands are fulfilled, those objective demands must be satisfied which are directly instrumental in the sustaining of the global developmental property of the entire structure. If these demands are neglected, then the danger arises that all practical social activities will come to a halt. This consideration supports the thesis of the primary role of the economic base in social development. Society as a structure must be continually sustained, and it is the function of many activities which constitute the social practice that will be called below the basic social practice (of production and exchange).

It should be noted that the Marxist sense of the term 'social practice' implies that effects contain the objective demand — those effects contain the objective demand towards which practical social activities are directed — such practical activities which are part of social practice and encompass that kind of social objective demand to which the practical social activities respond.

A given type of social practice includes such sets of objectively homogeneous, practical social activities which are relativized to effects of the types B, B', B'', etc. These effects comprise one and the same kind of objective demand. A type of social practice is also characterized by the type of all relevant objective conditions. For instance, all types of the basic social practice (concerned with production and exchange) are relativized to the conditions which determine the current level of development of the productive forces and the relations of production and exchange. These conditions will also be called basic conditions.

Therefore, a given type of (basic) social practice is a family of sets containing objectively homogeneous, practical social activities which are relativized to a type of effects which comprise the same kind of objective demand, and to objective conditions of a given type (especially of the basic type). Now, every element of the family, i.e., a set of objectively homogeneous, practical social activities which are partly ordered by the relation of temporal succession will be called a historical variant of a social practice of a given type.

In particular, one type of the basic social practice is exchange. It is determined by the same type of objective demand: acquisition of commodities produced by somebody else. This demand is an invariant kind of effect produced by practical social activities. It does not matter what the subjective social context of these activities happens to be. The sense of the activity is accomplished if the effects are produced, notwithstanding the differences that can be detected in historical variants of the practice of exchange, such as barter or exchange with a general equivalent.

We can say now that historically possible for a given society at a given time are the following variants: the current, historical variant of the (basic) social practice of a given type, and those potential variants for which there is an objective demand and which possess conceivable correlates in the form of single practical activities. Each of the historically possible variants is contained in the scope of the concept of historical possibility (in Marxist sense of the term). Owing to the fact that historical possibilities either continue to provide or begin to provide alternative answers to an objective demand, they either already possess effective, subjective social contexts or such contexts begin to emerge. In other words, every historical possibility either does or will contain practical social activities that are effective.

To come back to our earlier example, at a certain moment of social historical development two historical possibilities co-existed insofar as exchange was concerned, the two were one type of social practice. The first variant was simple exchange, a continuation of what existed before; the second variant was new, and involved exchange with a general equivalent. How did it happen that the second variant changed into a historical necessity?

The historical variant of exchange, i.e., barter, at some point could not solve the burgeoning contradiction, says Marx. The contradiction was: '(...) inherent in the commodity as such, namely that of being a particular use-value and simultaneously universal equivalent, and hence, a use-value for everybody or a universal use-value.' [8] We must deal with this issue at some length. The contradiction has been generated by two properties of the social relation called the relation of exchange. The first property resides in the commodity by virtue of its being exchanged as concrete use-values needed by the recipient. The second property resides in it by virtue of its being exchanged according to value, i.e., according to the socially indispensable amount of time needed for its production, i.e., independently of its use-value. These two properties are not logically contradictory in an obvious manner if only because they have co-existed for some time. Their co-existence, however, which is part of the objective conditions of that historical variant of exchange at least which is identified as simple exchange, is responsible for the quantitative growth of two series of objective effects : on the one hand, the objective demand for somebody else's commodities — somebody else's in the sense of not having been produced by the prospective user — spreads out, strengthens and becomes differentiated; on the other hand, the relation of simple exchange gains force and reproduces itself. At a certain stage of the development of these two series of objective effects, simple exchange ceases to be an answer to the objective demand; practical social activities which constitute it are no longer effective. A further development of simple exchange is blocked by a dialectic contradiction — the historical possibility of exchange with a general equivalent becomes a necessity. Only this possibility is a 'solution' of the encountered dialectical contradiction.[9]

Let us try to make this example a bit more general. Social relations which are part of the (basic) objective conditions of the (basic) social practice of a given type possess pairs of properties owing to which every historical variant of the practice of that type leads, as it continues to function, toward a quantitative growth of two series of objective effects. On the one hand, this kind of objective demand to which this variant responds takes roots and multiplies its forms; on the other hand, social relations which are part of the objective conditions of the historical variant of that practice become stronger. The relationship between the properties which make up a pair will be called a *dialectical contradiction* characteristic for a given type of (basic) social practice. Taken together, both series of effects of a specific historical variant of social practice of a given type will be called the global effect of the practice.

The foregoing remarks make it clear that development of social practice of a given type consists in the transition from one historical variant — when it has been blocked or is impeded by a dialectical contradiction — to the next, whose developmental prospects are open. In other words, development of social practices of different (basic) types [10] consists in persistent solving of the contradictions that they breed.

Thus, among historically possible variants of the (basic) social practice of a given type, historically necessary is that variant which in the further social development remains viable as the only variant that responds to the objective demand, and provides the only solution of the dialectical contradictions characteristic for the current type of social practice.[11] I will call this variant historical necessity (and take this concept to be the *explicans* of the Marxist understanding of that term).

In the framework of this interpretation of Marxist concepts of historical possibility and necessity, the concepts are referred to historical variants of a social practice of a given type in a given society at a given time. As matter of fact, however, the range of these concepts is broader. The successive variants of social practice are connected with correlated subjective social context, and therefore, the historical development of the social practice is at the same time a development of these contexts, i.e., of the correlated forms of social consciousness. On the other hand, substitution of one variant of social practice by another is connected with a change of the attending objective conditions which are formed and reproduced by each variant.

Consequently, historical possibilities and necessities of social practice determine developmental possibilities and necessities of the objective conditions and of social consciousness.

NOTES

[1] W. Dray, *Laws and Explanation in History*, Oxford, 1957, pp. 157, 161.
[2] J. Kmita, *Z metodologicznych problemów interpretacji humanistycznej* (*Methodological Problems in Humanistic Interpretation*), Warszawa, 1971, p. 28.
[3] F. Engels, *Ludwig Feuerbach and the End of Classical German Philosophy*, in: K. Marx and F. Engels, *Selected Works in Three Volumes*, Moscow, 1977, Vol. 3, p. 352.

[4] K. Marx, 'Preface to *A Contribution to the Critique of Political Economy*', in: K. Marx and F. Engels, cited above, Vol. 1, p. 504.

[5] K. Marx, *Capital*, Moscow, 1954, p. 48.

[6] For the sake of simplicity, I omit the intermediate variant which is connected with the developed form of value, but still coupled with a 'particular equivalent'. Cf. K. Marx, *Capital*, cited above, pp. 62–69.

[7] K. Marx, *A Contribution to the Critique of Political Economy*, in: K. Marx and F. Engels, cited above, Vol. 1, p. 50.

[8] *Ibid.*, p. 48.

[9] Cf. *ibid.*, pp. 49–51.

[10] Along with the basic type of social practice, other types, functionally subordinated to the basic practice, should be enumerated like scientific practice or artistic practice. It is impossible to give a detailed characterization of these practices in this essay. I want only to emphasize that the remaining types of social practice respond indirectly to the demands arising in the basic social practice. Compare articles by A. Pałubicka, 'Nauka i doświadczenie społeczne' (Science and Social Practical Knowledge), *Nurt*, 1973, No. 3, and 'Teoria naukowa i doświadczenie społeczne' (Scientific Theory and Social Practical Knowledge), *Nurt*, 1974, No. 4. The development of social practice is presented in these articles in the way that I want to characterize here in detail, as a continuous process of solving contradictions that are specific to a given type of practice.

[11] Historical development of social practice of any kind can be presented in two ways, by taking into account only the necessary, historically successive variants of the practice, or by noticing also the parallel variants until the moment they have been blocked by a relevant dialectical contradiction. The first approach produces a picture of development that can be metaphorically called 'the path of necessity'. Engels called it 'the logical method', and the second approach he called 'the historical method'. '(...) History often moves in leaps and bounds and zigzags, and as this would have to be followed through-out, it would mean not only that a considerable amount of material of slight importance would have to be absorbed, but also that the train of thought would frequently have to be interrupted (...). The logical method of approach was therefore the only suitable one. This, however, is indeed nothing but the historical method, only stripped of the historical form and of interfering contingencies.' F. Engels, *K. Marx — A Contribution to the Critique of Political Economy*, in: K. Marx and F. Engels, cited above, Vol. 1, pp. 513–514.

THE CONTROVERSY ABOUT THE DETERMINANTS OF THE GROWTH OF SCIENCE

3.1. SCIENCE AS A TYPE OF SOCIAL PRACTICE

As it is well known, there are two basic ways of understanding the term 'science' in epistemology. On the one hand, the concept is used to refer to a system of investigative activities; on the other, it is referred to a set of assumptions of scientific research and to its results expressed in sentences. To avoid a possible misunderstanding occasioned by this ambiguity, in the first instance I will use the phrase 'research (or scientific) practice', in the second, I will use the term 'science'.

Starting from the assumption that scientific practice is one of the types of social practice, that it is one of the 'autonomous segments' of the social division of labour, and with the help of the findings made in the previous essay, we are going to proceed toward some noteworthy conclusions about the subject of epistmological reflection.

Thus, first of all, the very fact that this sort of practice can be contrasted with others (of course, not before a certain level of social historical development has been reached) is an indication of a theoretical possibility and need to propose such characterization of social practice in which scientific practice stands apart from other forms of social practice. This characterization should comprise an elucidation of the objective function of the scientific practice in contradistinction to the remaining types of social practice, and especially to the basic practice. The latter has a distinct character which defines it in relation to all other types of social practice and indicates that its development functionally determines the development of other practices. On the other hand, we must bear in mind that a definition of the functional operation of the scientific practice with regard to the basic social practice must have a framework character to allow for changes through history. The function of scientific practice has been different in different stages of its development, and consequently it has also been different for various fields of scientific practice in the same period of time, as different disciplines of the same chronological period can, and as a rule do, represent different developmental stages. However, in different developmental stages of the scientific practice which functions differently in each of these stages, certain invariants of function must be found, or else there would be no reason to speak of them all as stages of the same type of social practice. These invariants must be identified by the framework characterization of the functioning mode of scientific practice.

It should be noted that different modes of objective functioning of various types of scientific practice can be discerned in the fact that they are more or less directly connected with the basic social practice. And so it seems that, e.g., the scientific practice connected with mathematical-natural disciplines is to a great extent directly connected with the basic social productive practice as far as its function is concerned, while in class societies, the humanist research practice is directly connected in the first place with the class biased political-*cum*-legal practice, and only through that practice with the basic practice.

Before I present my proposal of a framework characterization of the objective function of the scientific social practice, I will mention that according to the findings made in the previous essay, every type of social practice is characterized by its own, historically changing, subjective social context. It comprises normative beliefs which determine evaluative orderings, and, in particular, the purposes of activities and beliefs which determine (either alone or together with other factors) the methods of achieving selected values, i.e., purposes. Beliefs of the latter kind will be called predictive. Although the term may not have been happily selected — because not every belief, taken in isolation, can be a basis of prediction — it is more appropriate than the plausible alternative, viz., 'descriptive beliefs', because the beliefs I have in mind may have an evaluative character [1] and may often be expressed in the verbal form of a directive (which as a matter of fact is often done) and such utterances are not normally considered descriptive statements.

I propose to define the objective function of the social scientific practice in the following manner: *it codifies and deductively systematizes predictive elements of the direct, subjective social context of the basic practice*, and of subjective contexts of the remaining types of social practice functionally subordinated to the basic practice. This definition, as proposed, will be henceforth considered as one of the crucial assumptions of the investigations made in this book.

Let us now turn to the question of the status of scientific practice as a separate type of social practice. Clearly, it has its own, specific objective demands responsible for its character. These demands emerge in the course of historical development of scientific practice, and are represented on the subjective-social plane by normative and predictive beliefs. Predictive beliefs are typically verbalized as methodological directives, and determine the way in which the most preferred cognitive values are achieved. A system of beliefs of these two kinds will be called a social methodological consciousness of scientific practice. It has characteristic forms in different historical periods, and, possibly, in different disciplines.

As scientific practice is directly or indirectly subordinated, as far as its function is concerned, to basic practice, objective demands of any scientific practice are derivable from the objective demands of the basic practice. It means that the former are constituted of such potential global effects of the scientific practice — expressed as systems of statements — without which objective demands of the basic practice, or of other types of social practice which lie between the two, could not find a sufficiently adequate answer.

At the same time, however, one should remember about the relative autonomy of scientific practice: if in a certain period of time specific objective demands were formed, having been 'enforced' by the functionally superior types of social practice — in the extreme case by the basic practice — subjectively represented by a specific methodological consciousness, then the dominant norms and directives, typical for that consciousness, may remain intact and may continue to (subjectively) regulate scientific practice before new objective demands for scientific practice have appeared and made their current counterparts obsolete, i.e., ineffective.

Finally, even in the same developmental stage of a scientific practice, in certain fields different variants may co-exist as possible co-variants responding in different ways to the same demand. Only in the course of their further development is the number of these possibilities reduced by their characteristic, internal dialectical contradictions.

By viewing science as a form of social consciousness functionally determined by social scientific practice, or by viewing science as a subjective social context of social scientific practice, we have identified two spheres in science: (i) the sphere which has been called social methodological consciousness, and (ii) the sphere of scientific findings, i.e., scientific theories or 'bodies of beliefs' connected more loosely. Research results which are contained in sphere (ii) are achieved on the subjective grounds of a specific, social methodological consciousness, although the order of objective determinations is reverse. The demand for a particular type of research results is objectively primitive whereas methodological consciousness is its subjective representation. If this subjective representation is adequate, which means that its correlated subjective practice is effective, this representation becomes a social, fairly stable subjective context of that practice, and for a longer period of time may function as a 'referee' for further research results.

Philosophical reflection on social methodological consciousness, which strives to reconstruct it, verbalize it, explain its development, etc. — the scope of research activities that are undertaken under this rubric depends on the philosophical orientation represented by the researcher — will be called a theory of scientific cognition. As I have said in the Introduction, the term 'methodology of science' could be used as well, but it is too ambiguous — as it is used as a name for the social methodological consciousness (or especially as a name of one of its parts, viz., the prescriptive part) or as a name for the result of philosophical verbalization of the social methodological consciousness. Sometimes it also stands for a discipline which has limited its scope of interest to the reconstruction of specific, historical forms of social methodological consciousness.[2] For these reasons, I am more content with the term 'theory of scientific cognition', and I will use it in the meaning that has just been expounded.

3.2. HYPOTHETISTIC MODEL OF THE GROWTH OF SCIENTIFIC KNOWLEDGE

Hypotheticism, whose contemporary variant has been codified and remarkably enlarged by K. R. Popper, belongs, in my opinion, to a group of theories of scientific cognition which represent more than a mere record of an individual philosophical position. The group represents a certain stage in the development of scientific practice; it expresses social methodological consciousness that corresponds to that practice. In this essay, however, it is not my intention to find an answer to the question about the character of that developmental stage or about the degree of adequacy with which it is expressed in hypotheticism. My task is much more modest. I want to present the hypothetistic point of view on the development of scientific knowledge, some of its internal difficulties, and then to compare it with the assumption proper (in my opinion) for the Marxist approach.

As far back as the time of his writing of *Logik der Forschung*, K. R. Popper formulated a characteristic answer to the question about the methodological character of the directives of science: 'The way in which one answers these questions will largely depend upon one's attitude to science. Those who, like the positivist, see empirical science as a system of statements which satisfy certain logical criteria, such as meaningfulness or verifiability, will give one answer. A very different answer will be given by those who tend to see (as I do) the distinguishing characterstic of empirical statements in their susceptibility to revision — in the fact that they can be criticized, and superseded by better ones; and who regard it as their task to analyse the characteristic ability of science to advance, and the characteristic manner in which a choice is made, in crucial cases, between conflicting systems of theories.' [3]

Thus, the author of *Logik der Forschung* reconstructs social methodological consciousness of scientific practice (of course, all I am saying is an interpretation of his research procedure) with the use of directives which serve not so much to identify the methods of obtaining statements with certain qualitative methodological properties as to determine how the received body of scientific knowledge should be questioned and replaced by better knowledge — better according to some comparative criteria. This circumstance has been highlighted by one of the most faithful followers of Popper, I. Lakatos: 'While neoclassical empiricism inherited from classical empiricism only the problem of a monolithic, all-purpose appraisal of hypotheses, Popper's critical empiricism focussed attention on the problem of their discovery. The Popperian scientist makes separate appraisals corresponding to the separate stages of discovery.' [4]

It can be easily noted that such a theory of scientific cognition must have a historical character in the special sense of the term, at least inasmuch as it is opposed to the position of the Vienna Circle. Unlike the philosophers of that school, who almost totally neglected diachronic question in their 'logical analysis of science' (which, in their terminology, corresponded to social methodological consciousness of scientific practice), K. R. Popper puts forward the question: How does scientific knowledge develop? I think that the essence of this *sui generis* historical approach is best expressed by I. Lakatos in his statement: 'Methodology is wedded to history, since methodology is nothing but a rational reconstruction of history of the growth of knowledge. Because of the imperfection of the scientist, some of the actual history is a caricature of its rational reconstruction; because of the imperfection of the methodologist, some methodologies are caricatures of actual history.' [5]

The hypothetistic theory of scientific cognition, in its Popperian version, verbalizes social methodological consciousness in such a way that its content must explain by itself the succession of several stages of scientific knowledge from its very beginning. Thus we are facing here a version of historicism in the treatment of science, conceived of as a body of research results, or more precisely, as sequences of these results, each of which supersedes its predecessors. This historism is based on ahistorism in the treatment of social methodological consciousness which is allegedly historically unchanging, as it must supposedly always respect the so-called principle of rational criticism whose application warrants a steady progress of scientific knowledge.

The fundamental cognitive value, the main objective to be achieved by this super-historical methodological consciousness postulated by K. R. Popper, is explanation of certain facts and regularities: '(...) it is the aim of science to find satisfactory explanation, of whatever strikes us as being in need of explanation. (...) the *explicandum* is more or less well known to be true, or assumed to be so known. (...) The *explicans*, on the other hand, which is the object of our search, will as a rule not be known: it will have to be discovered.' [6] To solve a problem of explanation, we propose a 'bold' hypothetical theory which can be considered a satisfactory *explanans* if it passes the test of 'severe criticism'. The measure of severity is the difference between the probability of getting a positive test result on the assumption that the adopted hypothesis is true and the probability of getting a positive result on the assumption that a competing — usually a received — theory is true. In the extreme and most telling case, the proposed hypothesis entails (possibly together with certain initial conditions) some base statements (in the sense defined by the author of *Logik der Forschung* [7]) incompatible with the currently accepted rival theory. Severity of criticism reaches a maximum : the first probability equals certainty, the second probability equals zero. In other words, the idea behind severe criticism is to find out such base statements which would be true from the point of view of the hypothetical theory (or highly probable on its grounds) and false from the point of view of the competing theory (or of low probability on its grounds). If it can be shown that the base statements are true, that the hypothetical theory has been 'corroborated', it can be accepted tentatively. If the base statements turn out to be false, the hypothesis has been falsified and must be rejected.

The principle which underlies the procedure of applying severe criticism I will call the *principle of rational criticism*. [8] All tentatively accepted hypothetical theories become falsified, sonner or later. Consequently, the growth of science follows the pattern: P_1—TT—EE—P_2. The starting point for every cycle is a problem to be explained, P_1. A solution proposed as its explanation is a tentative theory, TT. It is made subject of critical analysis, or severe criticism, in the phase of error elimination, EE. The falsified theory is rejected and a new problem to be explained shows up, P_2, which initiates a new cycle in the same pattern.[9]

It must be pointed out that this Popperian pattern of the growth of 'objective knowledge', i.e., scientific knowledge, belongs to the 'third world', which is a specific interpretation of the idealist 'objective spirit': '(...) the third word, the world of objective knowledge (or more generally of the objective spirit) is man-made. But it is to be stressed that this world exists to a large extent autonomously; that it generates its own problems, especially those connected with methods of growth; and that its impact on any one of us, even on the most original of creative thinkers, vastly exceeds the impact which any of us can make upon it.'[10] The Popperian 'third world' is distinguished from physical phenomena on the one hand, and the realm of individual mental phenomena on the other. To be sure, they are all interconnected, directly and indirectly. The 'third world' affects the 'first world', or the physical world, through the 'second world', the world of individual mental events. These events in turn are in direct interaction with the 'third world' and the 'first world'.

It has been often remarked that the conception of the 'third world' formulated by K. R. Popper in the sixties is something entirely new in his philosophy, or that it is an addition he had to devise in order to have a province in which the principle of rational criticism could reign — permanently violated in the 'second world', factual and irrational domain as it is. I disagree with both these opinions. First, the conception of the 'third world' was implicit as far back as the time of publishing of *Logik der Forschung*, where 'objective knowledge' was contrasted with 'subjective knowledge of ourselves' in order to draw a line between base statements included among the tenets of 'objective knowledge' as intersubjective claims accepted by scientists and the statements which are reports on sense data coming from sundry individuals, which provide a psychological motive for acceptance of certain base statements, but do not provide a justification of these statements. A description of these perceptual findings belongs to the 'knowledge of ourselves', but does not belong to 'objective knowledge'. Consequently, the conception of the 'third world' is hardly a surprising part of the philosophical views of the author of *Logik der Forschung*, and one can seriously doubt that it has been created only in order to have a province ruled by the principle of rational criticism.

Any residual doubt may be dispelled by two further arguments. First, K. R. Popper never said that the 'second world' was irrational', i.e., that individuals violate the principle of rational criticism. On the contrary, he sided with Hume on the question of the right solution of the 'logical problem of induction'. It means that he agreed with Hume that induction does not justify beliefs, or is not respected in the 'third world of objective knowledge'. But he disagreed with Hume about the 'psychological solution of the problem of induction'. He did not accept the belief that individuals apply induction when accumulating their knowledge of the world. K. R. Popper thought that we learn individually in the same way as we develop our 'objective knowledge' in the 'third world', i.e., by elimination of the initial, false assumptions. Thus, contrary to Hume's opinion, we are not 'irrational' in the 'second world' (by 'believing in something that cannot be logically justified').[11] Secondly, if the 'third world', its autonomy notwithstanding, is a joint creation of several individuals, then, on the grounds of methodological individualism assumed by the author of *The Poverty of Historicism*, the 'rational' process of the growth of scientific knowledge could not be a result of an 'irrational' action of individuals.

The necessity to reconcile methodological individualism (that will be further discussed below) with the conception of the 'third world' led Popper to a position different from the ingenuous, psychological individualism of the positivists.[12] For the latter, the conception of social consciousness was acceptable only as a notion which denotes a set of beliefs commonly shared by all members of a given class of individuals, or possibly by a statistical majority of that class. For Popper, the 'third world' was a specific modified rendering of social consciousness — a much more complex construction. It was an area, so to speak, of intersubjective communication between individual subjects, especially the sentient ones, on which they shift from the individual grounds, where they confront their thoughts, to the common ground, where they compare the results of those confrontations — something like addition of vectors in a field of forces. As a physical force, when we speak literally about it, can be computed by an operation of adding up the contributing vectors whose joint effect is manifested in the movement of the object in a particular direction and with particular velocity — though it is not identical with any one of them — 'objective knowledge', which belongs to the 'third world', is a specific result of the action of individual minds. These minds are subjected to some developmental principles which can be formulated independently, but which depend, for the manner of their operation that has been metaphorically compared with addition of forces, on the principles of intelectual activity characterizing various individuals. In this way, the main assumption of methodological individualism, viz., the claim that social phenomena (in this case, phenomena concerning social consciousness) can be reduced to singular phenomena characteristic of sundry individuals (in this case, phenomena concerning their private consciousness), can be sustained.

But never mind the degree in which phenomena concerning individual minds determine processes in the 'third world' in the Popperian model of the growth of science, it is obvious that this model is a manifestation of a style of thinking characteristic for, as Engels called him, 'the historical ideologist'. 'The historical ideologist (...) possesses (...) material which has formed itself independently out of the thought of previous generations and has gone through its own independent course of development in the brains of these successive generations. True, external facts belonging to one or another sphere may have exercised a codetermining influence on this development, but the tacit presupposition is that these facts themselves are also only the fruits of a process of thought, and so we still remain within that realm of mere thought, which apparently has successfully digested even the hardest facts.' [13] In the Popperian model, research practice has been isolated from other types of social practice and is presented as exclusively determined by ahistorically defined set of norms and methodological directives which constitute an immutable standard of 'rationality' expressed in the principle of rational criticism.

A general characterization of the growth of scientific knowledge proffered by K. R. Popper has been the common starting point for the epistemological reflection of both adherents and opponents of hypotheticism. In some instances, the resulting views differ only slightly from Popper's philosophy. A case in point is philosophy of I. Lakatos, who proposed a slightly modified version of the conception of rational criticism by complementing Popper's definitioin of the term with a caveat: a falsified theory may remain within the boundaries of scientific knowledge if the theory which has unseated it does not have a 'corroborative surplus.' [14] In other instances, the Popperian point of view has been entirely demolished, especially in the philosophy of P. K. Feyerabend, or was used as a negative reference point, as in case of T. S. Kuhn. Nevertheless, it is remarkable that the controversy concerning the principles of the growth of scientific knowledge unfolds on the same grounds on which the Popperian model was built, namely on the plane of 'historical ideology', in the Engels' sense of the term.

3.3. WITHIN THE PURVIEW OF 'HISTORICAL IDEOLOGY'

One of the main premises of P. K. Feyerabend's philosophy [15] is a hypothetistic idea of the dependence of observation results on a presupposed theory: '(...) observations, and even more so observation statements and statements of experimental results, are always interpretation of the facts observed ; (...) they are interpretations in the light of theories. This is one of the main reason why it is always deceptively easy to find verifications of a theory, and why we have to adopt a highly critical attitude towards our theories if we do not wish to argue in circles: the attitude of trying to falsify them.' [16]

If we see the thesis of P. K. Feyerabend that observation is always impregnated by a theory all through its consequences, it will be evident that it cannot be reconciled, contrary to the opinion of Popper just quoted, with another fundamental idea of hypotheticism, namely that of severe criticism. Let us suppose, for instance, that we make subject of such criticism a hypothetical theory T and try to find such observable facts which should occur on the strength of a competing theory T', and which, if they actually occurred, would falsify theory T as incompatible with facts (and simultaneously corroborate theory T'). Naturally, the facts we would be looking for would have to be observable facts F' interpretable and susceptible to explanations offered by theory T'. Now we must decide what role to give to theory T' in the articulation of the observable facts F' that need to be explained. What in the description of these facts has had a purely observational, non-theoretical character? In other words: What could we have ascertained from observation if we had not had theory T' at our disposal, nor any other theory for that matter? Feyerabend says that we could have ascertained nothing, no definite state of affairs. 'Pure' observation can be expressed only in verbal reactions or non-verbal responses when they have no objective reference and are only causally connected with the observed states of affairs. Something like that happens when we react with the same utterances or gestures to states of affairs of the same kind (in some respect). We need a theory which is implicit in a language in which we register observable facts, to give our verbal reactions objective reference by making states of affairs spoken about in language stand for observable facts. If so much is granted, and there are many statements in Popper's writings that can be interpreted in this light, then the program of severe criticism falls through. As a rule, 'pure observational material' articulated by theory T' as observable facts F' and ostensibly lending itself to be used as severe criticism of theory T can be reinterpreted by T as a set of observable facts F which are compatible with T and explained by T. Thus T cannot be falsified.

Consequently, the program of severe criticism which was intended as a description of an actual practice stimulating growth of scientific knowledge, and which implied that old theories are to be replaced by new ones which have falsified them has been transformed by Feyerabend into a program of toleration. Many competing theories should be retained and allowed to process raw observational material according to their own standards. The main arguments in favour of that program are at the same time arguments against practical applicability of the principle of rational criticism.

It must be pointed out that the analysis of Feyerabend's destruction of the hypothetistic model of the growth of scientific knowledge leads to the conclusion, which he himself has overlooked, or perhaps was not been able to perceive in the framework of his views, namely that the concept of experience as an effect of observational contact between sundry individuals (or so many human organisms) and the world around them cannot play the role with which it was traditionally entrusted in epistemology — the effects of that contact cannot be the criterion of objective adequacy of a scientific theory. As a matter of fact, Popper has taken some steps away from that traditional position and eliminated the psychological notion of experience when he decided that base statements could not be justified by sense data registered by individuals. But he made no positive counterproposal which might suit his concept of experience corresponding to the thesis that base statements are adopted due to their intersubjective validity and not because they are vindicated by sense data. We will return to this question in the next essay.

Obviously, P. K. Feyerabend does not deny the historical

fact that various theories have been eliminated from the domain of scientific knowledge; he only indicates that they have been relegated therefrom by other forces than the principle of rational criticism. But aside from his critical contribution — to be found in his objection to the hypothetistic model of the growth of scientific knowledge, his emphasis on the 'empirical incomparability' of theories, and consequently in the claim that the program of severe criticism of hypotheses is unworkable — he offered a number of penetrating historical analyses of the succession of scientific theories, and pointed out, very convincingly, I believe, that these theories are incomparable also on the purely logical plane — by virtue of their syntactic and semantic properties. Statements which belong to two different theories of such kind, even if they are believed to be equivalent because they contain superficially identical terms, may have incomparable semantical objective references, a fact to be explained in the simplest case by the observation that these terms denote objects that belong to different set-theoretical categories. For instance, the notions of mass, force, acceleration, etc., denote properties (or sets of properties) of individual physical objects in classical physics, whereas the same notions denote relations (between same objects and a selected system of reference) in Einstein's physics. If Feyerabend's arguments are valid, then he has pointed out that it is impossible to demonstrate growth of knowledge not only in terms of rational criticism but, generally, in any logical terms. There is no sequence od systems of knowledge which remain in some stable logical relations to one another; they do not follow one another according to some principle which says that a new system replaces an old one if it stands in a certain logical relation to it.

On the other hand, I must point out again that P. K. Feyerabend does not deny that new theoretical systems are related in some way to the previous ones. For instance, relativistic mechanics is related to classical mechanics, and moreover, it is somehow constructed upon it, as Feyerabend admits.

But as he can find no purely logical relation between them —
no such relation is postulated by the principle of rational
criticism which could explain the succession of scientific systems, for instance — he tries to find a compromise between
a 'rational' and 'irrational' approach to the growth of knowledge. The 'rational' approach assumes that researchers systematically apply principles formulated in logical-*cum*-empirical
terms and rely exclusively on them when they compare scientific theories. The 'irrational' approach presupposes that researchers are moved by other motives that have nothing to do
with cognitive values when they choose between different
theories.

It is clear therefore that P. K. Feyerabend also stands on
the grounds of 'historical ideology', where the principle reigns
which stipulates that every thought has been elicited by another
thought, or more specifically, a thought which is constitutive
for a scientific theory or accepts one has been elicited by a
prior thought which endorsed certain 'rational' principles of
procedure or some aesthetic, religious, metaphysical, etc., guide-
lines which decide why a certain theory is more valuable
than another. Popper's 'historical ideology' is nevertheless
different from Feyerabend's 'historical ideology' in that, first,
the former favours a 'rational' approach — insofar as it postulates that the thought which elicits another thought does so
on the strength of a 'logical-empirical' principle that is used
to compare theories (namely, the principle of rational criticism) — and secondly, the determining thoughts are members
of the 'third world' as specific, systematically regulated joint
results of the efforts of individual intellects. P. K. Feyerabend's approach is more psychologistic.

The conclusions of Feyerabend's inquiries deserve to be
carefully examined, even if they do not contain any acceptable
positive solutions, because, in a manner unintended by himself, they show what roads of possible progress in the theory
of scientific cognition are closed. I have mentioned before his
analysis of the relationship between experience and theory in

which he demonstrated utter uselessness of the psychological concept of experience. Now I would like to focus on the thesis of 'logical-empirical' incomparability of certain contiguous stages in the development of scientific knowledge. This thesis is amply justified, I think, and the fact it states can be explained in a manner which accounts for the state of affairs ostensibly incompatible with that fact (at least from the perspective of 'rational historical ideology'), i.e., that successive stages in the development of scientific knowledge are somehow rested on its earlier stages.

It should be borne in mind that the results of the research practice of a given historical period, and especially scientific theories, are equipped with a certain social-humanist coefficient — to make use of a concept introduced by F. Znaniecki. This is to say that these results are cultural objects of some sort [17] if we consider the historically localized sphere of social consciousness which represent the social context of research practice in a particular period of time. This context consists of the current social-methodological consciousness and the research findings that have been accepted by it. From now on, I will speak in this sense of a methodological-theoretical humanist coefficient of a given theory, or alternatively, of a 'loose' body of research results. Apart from this historically localized humanist coefficient, a scientific theory does not exist as a social cultural object.

Social-methodological consciousness undergoes developmental changes, as I said in an earlier part of this essay. The methodological-theoretical coefficients of successive theories are often radically different. Consequently, if one accepts the possibility of a universal, perennial 'logical-empirical' comparability of different theories, it is usually an unconscious manifestation of an ahistorical position in the treatment of the methodological consciousness. Certainly, one cannot analyse an articulated scientific theory without looking at it from the point of view of a certain methodological-theoretical humanist coefficient, as no object deprived of an ambient social-

humanist coefficient can subsist as a work of art, no sequence
of sounds or inscriptions can subsist as a statement in language,
etc. Therefore, those who do not realize that a scientific theory
is a cultural object due to its shaping by an appropriate humanist coefficient, or even worse, those who do not realize
that this coefficient is constituted by a historically changing
sphere of social consciousness, take the norms and directives
which they accept to be an absolute yardstick, even if it is no
more than a subject of private acceptance. That was the case of
K. R. Popper. He encountered a number of difficulties when
he tried to show, using examples of concrete, historical
theories which succeeded one another, that, in his view,
the principal and perennial objective of science, i.e., explanation, is always achieved. He tried to evade these difficulties,
however, and said, for instance, that Newton's theory explains
Galileo's theory in its 'amended' version. The question is not,
however, what is the relationship between the 'amended'
theory of Galileo and Newton's theory, because the 'amended'
theory is already part of the theory which has corrected it,
but what is the relationship of the actual Galileo's theory,
taken in its authentic, original methodological-theoretical
coefficient, to Newton's theory, or what is the relationship
of the actual Galilei's theory to its version 'amended' à la
Newton. The latter relationships cannot be described exclusively through 'logical-empirical' comparisons if the methodological-theoretical humanist coefficients of such theories are
different. Incomparability of theories in pure 'logical-empirical' terms becomes even more evident as the historical analysis of them is done more meticulously. The conclusions of
P. K. Feyerabend referring to some physical theories which
date back to the turning points in the history of that discipline
show that the theories are incomparable in 'logical-empirical'
terms, and this result should surprise nobody.

In light of these remarks, it should neither be surprising
that naive attempts to apply formal logic to the analysis of
natural language end up in certain typical failures. I am speak-

ing especially of the (ultimately illusory) paradox of intensionality, which is closely connected with the question just discussed. Let us make a counterfactual assumption for the sake
of simplicity that vocabulary, grammar and the rules of deductive inference in natural language undergo no historical
changes, and let us concentrate on other factors which may
cause a modification of the rules of the objective reference
of the expressions in that language. All these factors will
ultimately boil down to one: to the fact that semantic rules
of objective reference are relativized to the actual system
of knowledge of the world. If certain expressions refer to
these things rather than some others, it is (1) because some
very general ontological assumptions have been made which
are responsible for splitting up all there is into categories
(possibly all these categories can be identified with set-theoretical ones), and (2) because some more specific synthetic
knowledge makes it so. If it is part of received knowledge
that 'All A's are B's', then the least one can say is that the
rules of objective reference must be such that the denotation
assigned to predicate A must be included in the denotation
assigned to predicate B. Or if it is part of received knowledge
that '$C(a, b)$', then the least one can say is that an ordered
pair of individual objects assigned by these rules to individual
terms a and b is such that the pair is an element of the denotation that these rules assign to the predicate C. To put it
briefly, objective reference of expressions of a given language
have a specific historically changing humanist coefficient
which has the form of an appropriate system of knowledge.

Let us assume now that there is a language L defined by
lexical rules, grammatical rules and rules of deduction, and
two successive, though not necessarily directly successive, humanist coefficients of semantic reference of expressions of
language L: systems of knowledge S_1 and S_2. It may well happen that a sentence of L is true (false) if its semantic reference is determined by S_1, and becomes false (true) if its
reference is determined by S_2. This is most obvious when we

speak in language L using objective reference determined by S_2 about beliefs of the people who expressed, or could have expressed, their beliefs in L from the point of view of S_1. Let us consider the (true) sentence: 'Władysław Jagiełło knew that Kraków lies on the Wisła'. Now, from the point of view of our knowledge which provides a humanist coefficient with objective reference for such phrases as 'Kraków' and 'the city whose one neighbourhood is called Nowa Huta', these two phrases refer to the same object. The same would not be true from the point of view of Władysław Jagiełło. It is not surprising therefore that the sentence: 'Władysław Jagiełło knew that the city whose one neighbourhood is called Nowa Huta lies on the Wisła', is false. It is because we have illegitimately substituted the humanist coefficient connected with the expression of our own beliefs in the Polish language for the humanist coefficient of objective references connected with expression of beliefs of Władysław Jagiełło in that language. Consequently, the change of the logical value of the entire sentence may seem paradoxical only to those who do not realize that there exist humanist coefficients of semantic references of the expressions of every natural language and that they change through history. I may add that the number of those who remain perplexed by the paradox of intensionality — and yet, stubbornly refuse to make account for specific humanist properties of subjectively relative concepts [18] which describe someone's beliefs through a characterization of the semantic references by relating them to the historically 'proper' humanist coefficient — is amazingly high.

As I have said above, the question of the so-called paradox of intensionality is closely connected with Feyerabend's thesis of incomparability of physical theories in 'logical-empirical' terms due to radically different methodological-theoretical humanist coefficients. It is connected because methodological-theoretical coefficients of these theories comprise humanist coefficients of semantic references of those expressions which have been used to formulate these theories. For that reason, in particular semantic references of these theories are in-comparable in 'logical-empirical' terms, and that means that the domains of objects described in these theories are incom-parable because they are relativized to different humanist mehodologically-theoretical coefficients. These domains, or the semantic references of the theories will be called their literal references.

To sum up the foregoing remarks, there are two types of conceptions of the growth of scientific knowledge which fall under the rubric of 'historical ideology'. One comprises con-ceptions that do not recognize the fact that phases of that growth are incomparable in 'logical-empirical' terms due to the variety of the literal references of the theories which correspond to these phases, or to put it more profoundly, due to the variety of the social methodological-theoretical hu-manist coefficients ; another type comprises such conceptions which recognize the existence that incomparability, but cannot identify its causes. Adherents of the first type of the conception of growth postulate an ahistoric acceptability of their own methodological-theoretical coefficient as a universal *tertium comparationis* of the successive theories, and in terms of that *tertium comparationis* formulate a principle which allegedly has been accepted either socially or privately by scientific researchers at all times. By the way, these conceptions are promoted in particular by certain philosophers who consider themselves Marxists — the normative part of the Marxist theory of scientific cognition is elevated by them to a system of perennial principles which are observed by the subjects of the scientific practice and constitute the ultimate determi-nant of the scientific growth. Conceptions of the second type are disarmed by the fact of the 'logical-emprical' incomparab-ility of the successive stages of the scientific knowledge. They surrender in the sense that they abandon, either partly or totally, the possibility of finding out 'rational' principles of the growth of science. They justify their decision by pointing to the interference from 'irrational' factors, or simply fail to vindicate their prostration altogether.

The approach of T. S. Kuhn is congenial to conceptions of the second type, even though, for obscure reasons, it is sometimes extolled as related to the Marvist theory of scientific cognition. It should be clear, however, that it is entirely encompassed within the framework of the 'historical ideology.'

The essence of T. S. Kuhn's opposition to the hypothetistic model of the growth of scientific knowledge can be presented as follows. New theories, connected with new phases in the growth of science, or speaking more precisely, with new 'paradigms' [19] containing new theories, neither appear in the history of science nor acquire wide recognition in the community of scientists as a result of falsification of the theories which ruled in the previous paradigm: 'Competition between segments of the scientific community is the only historical process that ever actually results in the rejection of one previously accepted theory or in the adoption of another.' [20] How are these 'segments' created? By the very fact that the same paradigms are accepted, i.e., the same patterns of scientific practice are endorsed comprising somehow associated together: laws, theories, their applications and technical equipment. History of science is thus a repetition of the same cycle comprising: (1) emergence and reinforcement of a new paradigm, (2) a period of its dominance, (3) a period of its decline, (4) emergence and reinforcement of a new paradigm.

We already know that none of the phases of this cycle follows its predecessor as a result of a critical argumentation which directly confronts two theories belonging to different paradigms. There is no principle of 'rational' succession. So what factors bring about a new phase? All that T. S. Kuhn has to say about it is that individual predilections and coincidences win the day. At some point, these predilections all cluster around one paradigm, and a new 'segment' is created which comprises a majority of the scientists of a given community who feel strong enough to impose their dictatorship. Thus, specifically, phase (1) begins when '(...) an individual or group (...) produces a synthesis able to attract most of the next generation's practitioners.' [21] When they form a sufficiently large percentage of their community, phase (2) sets in, the phase of dictatorship by a new paradigm: 'Those unwilling or unable to accomodate their work to it must proceed in isolation or attach themselves to some other group.' [22] Under the terror unleashed by the proponents of the new paradigm, who specifically control the process of education of the new acolytes in a given field of knowledge, phase (2) persists in spite of the fact that the paradigm is systematically challanged by phenomena that it patently cannot handle, a glaring fact perceived at first glance. These phenomena are called 'puzzles', and are either 'solved' within the paradigm in a strained fashion, or are counted among 'anomalies'. It is now necessary that by coincidence a large number of such anomalies be observed if phase (3) is to follow, the phase of 'crisis', which lasts until a new paradigm has emerged and the old one has been eliminated in a turmoil of a revolution (phase 4): '(...) scientific revolutions are inaugurated by a growing sense, (...) often restricted to a narrow subdivision of the scientific community, that an existing paradigm has ceased to function adequately in the exploration of an aspect of nature to which that paradigm itself had previously led the way.' [23]

In point of fact, T. S. Kuhn's greatest achievement is the introduction of his own term 'paradigm' for the methodological-theoretical humanist coefficient of the successive stages in the development of scientific knowledge and highlighting its role as social-subjective basis in the process of crystallization of the standards of 'scientific validity'. Aside from that, he characterizes its origins in a manner totally conformable with the point of view of 'historical ideology' and methodological individualism. A paradigm takes shape and sets in the social consciousness when it has been accepted by a majority of scientists motivated by various individual circumstances: metaphysical views, episodes in biography or idiosyncrasies of personality, aesthetic preferences, etc. Hence '(...) as that goes on, if the paradigm is one destined to win fight, the number and strength of the persuasive arguments in its favor will increase. More scientists will then be converted, and the exploration of the new paradigm will go on.' [24]

3.4. THE PRINCIPLE OF THE GROWTH OF SCIENCE

As I have mentioned above, an attempt to explain the fact registered by Feyerabend's thesis that stages in the development of scientific knowledge are incomparable opens a possibility of discovering a plane on which these stages can be tied together in some way. This option is satisfied by Marxist theory of scientific cognition as it contains premises for an answer to the question: How are social, methodological-theoretical humanist coefficients formed in various stages of the development of scientific knowledge? These stages remain incomparable only as long as we are locked within the purview of 'historical ideology'.

Among the remarks which opened this essay, there was one which said that it is the social function of the research practice to codify and systematically deduce predictive elements of the subjective social context for particular types of practice, and especially, for the basic practice. Now the system of objective conditions of such practice with which a social subjective context is connected, after it has been codified and deductively systematized by a given theory T, will be called a practical-objective reference of theory T. The literal reference of the latter constitutes a social-subjective manifestation of its practical-objective references. In view of my opening assumptions, the following relation obtains between the two references of theory T. An objective demand from a social practice of a given type is addressed to the scientific practice, and requires that it produces a theory of particular kind, that as a minimum can codify and deductively systematize appropriate predictive elements of the subjective social context of the next developmental phase of that practice. Which elements of that context are already there, and which have to be devised by the research practice, is a secondary question. It will be discussed in the next essay. But at least so much is clear that theory T proposed by research practice is accepted as a sufficiently adequate response to the objective demand if it functions effectively in the desirable way. It is then adopted together with norms and methodological directives which constitute the subjective context of the practice of its creation and represents subjectively the objective demand that has been directed to the research practice. The system of objective conditions of the developmental stage of a social practice of a given type, subjectively initiated by theory T (i.e., the practical-objective reference), explains why it is effective, and why theory T has been socially accepted. It explains the fact of social acceptance of its literal reference, i.e., of the scientific 'picture' of the appropriate fragment of nature, relativized to the humanist methodological-theoretical coefficient connected with theory T.

According to the thesis about relative social autonomy of research practice, when the social methodological-theoretical humanist coefficient has been established in the conditions described above (in Kuhn's language, a paradigm has emerged) it begins to operate autonomously as a criterion of scientific acceptability of new theories. Theories arise within its purview and, unlike theory T, in no more than an indirect connection with the demands of the social practice whose demands they are supposed to satisfy. This state of affairs persists until a new objective demand for a theory arises which cannot be satisfied by the existent methodological-theoretical coefficient. Then, sooner or later, a new theory is formed, say, T', in the same way in which T has been brought to being. Thus T' is a 'guide' for a new social methodological consciousness, for a new methodological-theoretical humanist coefficient. Just because this coefficient is new, T' is incomparable, in Feyerabend's sense of the term, with theory T.

The mechanism of social creation, or social acceptance, of the two theories, T and T', as outlined above, must be additionally described from the point of view of historical materialism by a very important condition. Social acceptance of either theory depends not only on a functional determinant, which has already been discussed, but also on a genetic determinant. The entire determination has therefore a double, genetic-functional nature. Theory T' cannot emerge in a mental vacuum. It must rely somehow (the genetic determinant) on the received 'though material', i.e., on theory T and several other theories which arose within the scope of operation of the same methodological-theoretical coefficient (for simplicity, I will call them all theory T). 'Here economy — says Engels — creates nothing anew, but it determines the way in which the thought material found in existence is altered and further developed, and that too for the most part indirectly.' [25]

Let us look more closely at the question just discussed. We can notice that theory T is eliminated by theory T' when it has become impossible for T to subjectively regulate variant V_1 of that domain of social practice which engages both theory T and (subsequently) T'. To the objective elimination of the earlier variant V_1 by the later variant V_2 corresponds subjective elimination of theory T by theory T'. Transition from V_1 to V_2 is not a historical necessity [26] as long as side by side with V_2 there exist other, equally viable and parallel options $V_3, ..., V_n$. It is a historical necessity in such circumstances that V_1 passes into an option in the alternative corresponding to the series $V_2, ..., V_n$. In a special case, the alternative may comprise only one option, that of V_2, and then a transition to V_2 is a historical necessity. Anyway, already in the variant V_1 objective conditions arise for a necessary transition to one of the permissible options (under these conditions) out of $V_2, ..., V_n$. Making use of the concept earlier introduced we can say that the practical-objective reference of theory T

historically passes into a domain with respect to which theory T is no longer adequate. The choice of the option out of various possibilities $V_2, ..., V_n$ may depend sometimes on the new theory proposed as a substitute for the old one.

Let us suppose that option V_2 wins out and is characterized by conditions which are the practical-objective reference of theory T'. Obviously, to say that the practical-objective reference of theory T' is identical with the system of the objective conditions of the historical variant V_2 of social practice means, in the light of the foregoing remarks, that that reference systematizes with sufficient efficiency the predictive knowledge which constitutes an immediate social subjective context of variant V_2, which is crucial for its gaining social acceptance and leads the way for the acceptance of the methodological-theoretical humanist coefficient 'guided' by theory T'. Theory T' possesses, naturally, its own literal reference which is incomparable with the literal reference of theory T if the methodological-theoretical coefficient of the former, as the subjective rendering of its practically-objective reference, is radically different from the respective coefficient of the latter.

We can ask the question now: How can the literal reference of theory T' be constituted without forfeiting the chance of an adequate substitution of its practical-objective reference? How can the latter be mapped out by the literal reference of theory T' as accurately as possible? In my opinion, *Capital* contains a practical answer to this question: the methodological-theoretical coefficient of theory T' should contain a norm which postulates that the construction of a theory should begin from a characterization of the practical-objective reference of the previous theory T, i.e., from a characterization of the objective conditions of variant V_1 of social practice, and then should proceed to analyse the mechanism responsible for the transformation of the former historical reference of the theory into a new system of objective conditions connected with the variant V_2. In this way, the approximation of the literal reference of theory T' to its practical-objective reference, i.e., to the system of objective conditions of variant V_2 of social practice, is an instance of the application of the method called by the founding fathers of Marxism, the *method of historical unfolding of concepts*. Let us note that the methodological norm which enjoins that this method be applied requires of theory T' that its literal objective reference comprises in the first place the practical-objective reference of the earlier theory T. If so much is granted, then theory T', constructed in compliance with this norm, (1) will contain the most important component of the functional-genetic *explanans* of the explanation of the social acceptance of theory T, because it will characterize the objective conditions of the practice effectively rationalized by theory T, and socially accepted for that reason, (2) will be inherently related to theory T and will treat its literal reference as an understandable (explainable) on the grounds of T' subjective expression of its practical-objective reference, and *a fortiori* as the proof of its own adequate characterization of that reference, (3) will become comparable due to its practical-objective, and not literal, reference.

Of course, I cannot put forward the thesis that the development of the social research practice is ruled by the methodological norm of the historical unfolding of concepts as a rival proposal to other approaches to the development of social research practice. I cannot do so for at least four reasons. First, such a proposal would imply that the development of science is decided by methodological norms accepted in the community of researchers. Second, it would imply that certain norms are perennial, and that the norm of historical unfolding of concepts is such. Third, the contention that the norm of historical unfolding of concepts was always accepted in the community of scholars would be simply false; this norm is no more than an important element of the normative aspect of Marxist theory of scientific cognition. Fourth, this norm, as every other norm of Marxist theory of scientific cognition, is derived from an appropriate statement which declares that it must historically come true. To elaborate a little bit more on the last reason, I can say that the methodological norm of historical unfolding of concepts, as a characteristic part of Marxism, relies on the observation of a certain developmental regularity connected with research practice. It shows a regularity which can be reconstructed within a framework description of the objective manner of functioning of that practice in all historical periods of its development. The norm of historical unfolding of concepts proposes such a practice of constructing theories which could guarantee that that regularity is displayed in the purest form, i.e., free of various interferences which impede its progress.

Taking up this point of view, I will try to characterize the regularity which underlies the methodological norm which requires first of all that concepts unfold through history. Then I will show that this regularity is a corollary of the fact that social research practice functions objectively with respect to overall social practice in the way that has been described above.

This regularity can be formulated as follows. The sufficient condition of social acceptance of a class of research results (theory) T' as a substitute of a previously accepted class of research results (theory) T is tantamount to simultaneous satisfaction of two conditions: (1) the literal reference of theory T' contains an approximation of the practical-objective reference of theory T which is at least as adequate as the literal reference of (theory) T; (2) T' codifies and deductively systematizes in an effective manner new elements of the direct subjective predictive social context of the successive (i.e., later than T) developmental phase of the appropriate type (types) of a non-scientific social practice. This sufficient condition will be called the *principle of the growth of science*. I will consider it the principal tenet of this essay and, at the same time, a counterproposal to other descriptions of the process of the development of scientific knowledge, a counterproposal made by Marxist epistemology.

Let us have a closer look at this principle. Condition (2) makes social acceptance of a given class of research results (theory) T' depend on the following circumstance: these results must effectively fulfil the framework objective function of any social practice, as it has been characterized in an earlier part of this essay, with respect to the successive developmental phase of the non-scientific social practice. In other words, they must be a sufficiently adequate answer to the current objective demand. The principle is therefore a consequence of the thesis about the functional component in the determination of social consciousness. This thesis interprets certain phenomena about social consciousness as an answer to objective demands of a relevant kind. Condition (1) represents the genetic aspect of the functional-genetic determination of the development of research practice. It must be examined under two headings, however, because, as a matter of fact, it is a summary description of two cases which form one alternative.

The first case obtains when T' is no more than a simple elaboration of T, and therefore T' contains T as its part. In this situation, the literal reference of the class of research results (theory) T' comprises the literal reference of the class of research results (theory) T. The literal reference of the former is therefore an equally adequate approximation to the practical-objective reference of T as the literal reference of (the class of research results) T. The second case corresponds to the situation of a 'cientific revolution': new research results which make up theory T' are connected with a radically new methodological-theoretical humanist coefficient. This situation occurs when the appropriate domain of research practice fulfils its objective function of codification and deductive systematization of the direct subjective predictive context of a non-scientific social practice of a relevant type in a historical moment of transition from one variant of that non-scientific practice to another. The context of the 'abdi-cating' variant loses its effectiveness, whereas the context of the 'accending' variant gains it by taking over all effective elements of its predecessor. The new context becomes now the object of the current scientific practice, which consequently produces a class of new research results, i.e. (theory) T', connected with a new methodological-theoretical humanist coefficient. I must emphasize now that in this situation the literal reference of T' contains a sub-domain which is a more adequate approximation to the practical-objective reference of T than the literal reference of T. It will be clear why it has to be so if we consider that, on the assumption which identifies an appropriate sub-domain of the literal reference of the class of results (theory) T' and the practical-objective reference of the class of results (theory) T, it is possible after adoption of appropriate premises to give a functional-genetic explanation of the social fact of acceptance of the class of results (theory) T. It is possible, in other words, to answer the question why was the knowledge expressed in T effective, in the context of a system of objective conditions represented by the reconstructed practical-objective reference of that knowledge — because its social acceptance was already decided by that effectiveness. It is impossible, however, to succeed in the attempt to proceed with explanation in the opposite direction.

Although it may seem at first glance that the regularity described above and concerning the development of the research practice, especially its condition (1), is a purely speculative postulate, acceptable as a mere result of the assumptions that have been adopted about the way in which research practice objectively functions against the background of the continually developing non-scientific practice — many facts established by historians of science indicate that this suspicion is ungrounded. It is a fact, for instance, that the literal reference of the heliocentric theory of Copernicus treated as a system of objective conditions for the domain of that social practice which makes use of astronomical knowledge (navigation, irrigation of fields, etc.) provides, with the help of additional premises, an explanation why astronomical knowledge contained and systematized by Ptolemy's theory had won social acceptance. It is possible, in other words, by considering objective conditions described by Copernicus' theory, to answer the question: Why was the knowledge contained and systematized in Ptolemy's theory practically effective? It is also a fact that in a similar manner it is possible to explain social acceptance of Galileo's physics, i.e., to explain effectiveness of the practice subjectively based on that physics, taking Newton's physics as a description of the objective conditions of that practice. The same can be said about Newton's physics viewed through Einsten's physics, or generally, about all such pairs of classes of research results (theories) in which the later element is a socially accepted continuation of the earlier element which was also socially accepted.

Moreover, it is also a fact that such continuation always contains as its component the class of earlier research results which either have or have not been modified, depending on whether the methodological-theoretical humanist coefficient connected with pertinent classes of research results remain in the relation described as the first or as the second case (cf. above), or, indirectly, and this is of utmost importance from the theoretical point of view, depending on the mutual relationship between historical variants of the relevant social practice with subjective, predictive social contexts which have been codified and systematized by the two classes of research results.

Finally, it is also a fact that the social methodological consciousness has registered in some way the developmental regularity of research practice, or, speaking more precisely, of the dependence (in the sense of the necessary condition) of social acceptance for the new set of research results (theory) on whether the literal reference of these results (theory) is not less adequate approximation to the practical-objective reference of the earlier class of research results than the literal reference of the latter. To be more specific, a methodological norm has appeared in physics, or in philosophical examination of the research practice of physics, called the *principle of correspondence*. It is formulated in different ways, most commonly as a demand that the earlier theory logically follows from the later one as its specific case, or, more cautiously, as its limiting case. This formulation is imprecise because it does not distinguish between two different relations that I have identified, in which successive theories may stand to one another. Consequently, it does not pay sufficient attention to the specificity of the relation of the second kind. If the relation between two theories is of the first kind, it is in fact true that the earlier theory logically follows from the later one — which is its simple expansion based on the same methodological-theoretical humanist coefficient. In the second case, when the new theory is not a logical premise for the earlier theory, it does not explain its predecessor. All that can be explained is the social acceptance of the earlier theory, when the appropriate fragment of the literal reference of the later theory is treated as the practical reference of the earlier theory (with the help of additional premises). This situation can be encapsulated in the statement, already formulated, that the later theory provides a more adequate approximation to the practical-objective reference of the earlier theory than the literal reference of the latter. It is, however, a delusion to believe that in this situation research results of the earlier theory logically follow from the research results of the later theory. The false impression that it might be so arises from the fact, already pointed out by me, that it is possible to derive statements from the later theory which correspond to statements belonging to the earlier theory. These modified counterparts have a literal reference only in the domain of the literal reference of the later theory, to which, as a matter of fact, they, still belong — analyses of Feyerabend and Kuhn fully support this belief. These references are at the same time practical-objective references of the authentic statements of the earlier theory, on the assumption that the later theory contains in the domain of its literal reference a sub-domain which is identical with practical-objective reference of the earlier theory. Thus, although it is customary to retain in the modified version of the statements belonging to the earlier theory, the same names which appear in the original formulation of the theory — and this custom may be justified by the fact that the original statements, or the earlier beliefs, are specifically utilized by the later theory — yet, when perpetuated uncritically, this custom is responsible for the deception just described. For instance, the law of free fall of Galileo, as formulated in Newton's physics, and later on in Einstein's physics, has had very different literal reference in the later context from what it had in the 'original' context. And conversely, Newton's or Einstein's physics can be used to derive only a modification of that 'original' law, and not the law itself. Such modifications are characterized by literal references which are, in the light of the foregoing arguments, ever more adequate approximations to the practical-objective reference of the law of free fall.

I would like to continue using the term 'principle of correspondence', as I share the intuitions which inspired its introduction, but I do not wish to perpetrate further deceptions spread by that term, and so I will give it a precise meaning. The principle of correspondence is the statement to the effect that it is the necessary condition of social acceptance of a class of research results (theory) T' as a substitute for the socially accepted class of research results (theory) T that the literal reference of (theory) T' contains an approximation to a practical-objective reference of theory T which is at least as adequate as the literal reference of (theory) T.[29]

It will be seen from this formulation of the principle of correspondence that, the way I see it, it makes social acceptance of a class of research results depend on the fulfilment of the first of the two conditions that were enumerated in my definition of the principle of the development of science. According to that principle, the condition listed in the principle of correspondence and the principle which requires that the new class of research results makes account of new facts in the sphere of social-subjective predictive context of a non-scientific social practice combine to produce the sufficient condition for the social acceptance of the new class of research results (a theory).

I will additionally define the directive of correspondence as a methodological directive which stipulates that a scientific theory be justified in such a way that its justification shows how the theory fulfils the formal condition imposed by the principle of correspondence. In light of the earlier postulates, the directive of correspondence requires that the new theory explain in the functional-genetic aspect the fact of social acceptance of the received theory (or, generally, of the earlier research results) on the assumption that the domain of the literal reference of the new theory comprises a sub-domain which is the practical-objective reference of the received theory (a class of received research results). Naturally, every new socially accepted theory satisfies the principle of correspondence, as it is interpreted here. However, a conscious application of the directive of correspondence, which leads to the most effective fulfilment of the principle of correspondence, depends on the acceptance of Marxist theory of scientific cognition in the form that it has received here. The directive of correspondence belongs to the normative part of that theory, and it is always satisfied if a new theory is constructed with the help of the method of historical unfolding of concepts.

NOTES

[1] Or 'evaluation', in the sense which I expounded in 'O niejednorodności aksjologicznej predykatów oceniających' (On Axiological Heterogeneity of Evaluative Predicates), *Studia Semiotyczne*, Vol. 4.

2 I was using the concept of methodology of sciences only a short time ago. Now I believe, however, that separation of a discipline which would be solely concerned with a reconstruction of the social methodological consciousness and leave the study of its development to another discipline (history of science) is not theoretically justifiable.

3 K. R. Popper, *The Logic of Scientific Discovery*, London, 1959, p. 49, 50.

4 I. Lakatos, 'Changes in the Problem of Inductive Logic', in: *The Problem of Inductive Logic*, ed. I. Lakatos, Amsterdam, 1968, p. 375. It may be opportune to mention that Lakatos calls logical positivism 'neoclassical empiricism'.

5 *Ibid.*, p. 387.

6 K. R. Popper. 'The Aim of Science', in: K. R. Popper, *Objective Knowledge, An Evolutionary Approach*, Oxford, 1973, p. 191.

7 K. R. Popper, *The Logic of Scientific Discovery*, cited above, pp. 100–104.

8 For that reason, reconstruction of the development of scientific knowledge based on the principle that this development meets the requirements of rational criticism is called by I. Lakatos, a rational reconstruction. In fact, 'rationality' does not belong to the reconstruction but to the development that it describes. I would like to underscore the fact that rationality so conceived has little to do with the assumption of rationality which I postulated in Z *metodologicznych problemów interpretacji humanistycznej (Methodological Problems in Humanistic Interpretation)*, Warszawa, 1971, as a tacitly adopted premise of humanist interpretation. In particular, this conception, unlike mine, does not make any substantive conditions for the beliefs of individuals, and especially it does not assume that these beliefs comprise, for instance, the principle of rational criticism. Note that the term 'critical rationalism' is now used ordinarily, but it seems to me that 'rational criticism' is more accountable.

9 Cf. K. R. Popper, 'Epistemology without a Knowing Subject', in: K. R. Popper, *Objective knowledge*, cited above, pp. 118–119.

10 K. R. Popper, cited above, p. 147.

11 Cf. K. R. Popper, 'Conjectural Knowledge: My Solution of the Problem of Induction', in K. R. Popper, cited above, pp. 1–31.

12 The psychological variant of methodological individualism has been criticized by him earlier in *The Open Society and Its Enemies*, London, 1947, Vol. 2, p. 87.

13 F. Engels, 'To F. Mehring', in: K. Marx and F. Engels, *Selected Works in Three Volumes*, Moscow, 1977, Vol. 3, p. 496.

14 Theory T has a corroborative surplus as compared with theory T' if a surplus of its empirical contents over the contents of theory T' has been corroborated, but not the other way round. The reservation of

I. Lakatos quoted above introduced the most important alteration to
the model of the growth of knowledge postulated by Popper. But it
has not been a fundamental change. Consequently, I cannot agree with
the opinion of S. Amsterdamski (in *Między doświadczeniem a metafi-
zyką (Between Experience and Methaphysics)*, Warszawa, 1973, p. 188)
that 'Nothing makes this view similar to the prime and fundamental
idea of Popper's except, perhaps, the belief that all statements have a
theoretical character, plus the program of building a logic of scientific
discovery.'
[15] I have mainly in mind the following essays by P. K. Feyerabend:
'Explanation, Reduction and Empiricism', in: H. Feigl and G. Maxwell,
Minnesota Studies in the Philosophy of Science, New York, 1962,
Vol. 2; 'How to Be a Good Empiricist — A Plea for Tolerance in Matters
Epistemological', in: B. Baumrin, *Philosophy of Science*, New York,
1963, Vol. 2; 'Science without Experience', in: *Jorunal of Philosophy*,
1969, No. 22.
[16] K. R. Popper, *The Logic of Scientific Discovery*, cited above, p. 107.
[17] I assume the point of view presented more broadly in my book,
Z metodologicznych problemów interpretacji humanistycznej, cited
above, pp. 34–38. Recently, I dealt with the same question in 'Czy artysta
jest twórcą dzieła sztuki ?' (Does the Artist Create the Work of Art ?),
in: *Sztuka*, 1974, No. 4.
[18] A broader characterization of humanist subjective concepts is to be
found in the book by K. Zamiara, *Metodologiczne znaczenie sporu o
status poznawczy teorii (The Methodological Significance of the Controversy over the Epistemological Status of Theories)*, Warszawa, 1974,
pp. 141–143.
[19] It will be easily perceived that Kuhn's 'paradigm' corresponds to
what I called the humanist methodological-theoretical coefficient of
a body of research results.
[20] T. S. Kuhn, *The Structure of Scientific Revolutions*, Chicago,
1965 p. 8.
[21] *Ibid.*, p. 18.
[22] *Ibid.*, p. 19.
[23] *Ibid.*, p. 91.
[24] *Ibid.*, p. 158.
[25] F. Engels, 'To K. Schmidt', in: K. Marx and F. Engels, cited above,
Vol. 3, p. 494. Engels' utterance which I have quoted refers to philosophy,
but it can be extended to cover all science. Let me also add that Engels'
'economy' comprises a set of objective conditions and the objective
function of the basic social practice.
[26] Cf. previous essay.
[27] There is, however, a domain of social research practice which codifies

and systematically deduces just the subjective social context of the research practice — the theory of scientific cognition. Also this theory is subject to the developmental regularity just described. To underscore descriptively this state of affairs, I would have had to complicate my formulations in the main body of the text, so I decided to transfer this comment to the footnote.

[28] This formulation of the principle of correspondence differs from the version which I have put forward in some of my recent publications (for instance, in the article written together with W. Ławniczak, 'Marksizm a empiryzm' (Marxism and Empiricism), *Studia Filozoficzne*, 1974, No. 10). A revised fragment of that article of my autorship is included in the next essay. The most important difference, if we put terminological alternations apart, is that in my previous writings I identified the principle of correspondence with the formula which is now called the principle of the growth of science.

THE RELATION OF MARXIST
EPISTEMOLOGY TO EMPIRICISM

In the previous essay, I have compared various contemporary conceptions of the growth of science and confronted them, in general terms, with the view typical for Marxist theory of scientific cognition, as I understand it. In this essay, I will undertake a similar comparison of various conceptions of empirical foundations of sciences. In both cases, what I have to say goes only marginally beyond the scope of purely immanent critique of the literal contents of various points of view. A full critical analysis of the 'thought material', especially of the material accumulated by traditional empiricism, could only be made after a detailed historical study has been completed and resulted in a presentation of the developmental phases of several branches of research practice, which have grown under the historically changing, objective influence of a non-scientific social practice on the one hand, and found their subjective expression in different variants of traditional empiricism on the other. These variants should then be analysed as philosophical verbalizations of successive developmental stages of the social methodological consciousness. Such historical studies, fully applying the directive of correspondence in theory of scientific cognition, are at the moment no more than a design for the future. But even in full awareness of its limitations, a confrontation of various elements of philosophical verbalization of different developmental stages of the social methodological consciousness can be useful as an attempt to systematize 'thought material' and present the uniqueness of the Marxist point of view.

I repeat, however, that such confrontation cannot provide sufficient grounds for a Maxist criticism of other doctrines. In conformance with what has been said above and with the approach that was outlined in the previous essay, I may say that it is possible to identify two levels of Marxist analysis of various doctrines: scientific-theoretical, philosophical, ideological, etc. The first, elementary level must be reached, and the work upon it must be completed before an analytic operation on the second level has begun. The first level involves a literal understanding of the contents of a given doctrine, and then produces a historical humanist intepretation of that doctrine rather than an adaptive interpretation. The second level, theoretically more profound, encompasses the totality of cognitive operations with the view to revealing the origins and function of the doctrine which has been literally reconstructed at the first level. Only then does it become important and feasible to ascertain the practically-objective reference of the doctrine and provide thereby an answer to the question: What system of social-historical objective conditions has found its subjective representation in that doctrine? It is obvious to anyone who has studied with some application the writings of the founding fathers of Marxism, and *Capital* especially, that these two levels of analysis are always present there. For obviously, if these authors took so much time to investigate the past and contemporary theoretical-philosophical conceptions, they certainly did not do so because they wanted to lay it out plainly that these conceptions were logically incompatible with the doctrine of their own and therefore had to be criticized.[1] They were preoccupied with something else: application of the directive of correspondence to, or a second-level analysis of, the origins and the social-historical function of the ideas they set out to exploit. In the process, they turned these ideas, developed undoubtedly in accordance with the directive of correspondence, into evidence which supported their own theoretical diagnoses concerning the domain of the objective conditions of social practice, the conditions which constituted the practical-objective reference of those ideas. It will be easily noticed that following that procedure they obtained, as if automatically, an evaluation of adequacy of the conception under analysis with respect to its practical-objective reference.

4.1. METHODOLOGICAL INDIVIDUALISM OF TRADITIONAL EMPIRICISM

The common denominator of all non-Marxist versions of empiricism is to be found in the conception called methodological individualism. The name was selected and expressly used by K. R. Popper with the intention of encapsulating therein the essence of his views as opposed to Mill's psychologism. In *The Open Society and Its Enemies*, Popper said: 'Psychologism is, I believe, correct only insofar as it insists upon what may be called "methodological individualism" as opposed to "methodological collectivism"; it rightly insists that the "behaviour and the "actions" of collectives, such as states or social groups, must be reduced to the behaviour and to the actions of human individuals.' [2] One of the distinguished adherents of Popper's conceptions and his follower, J. W. N. Watkins, formulated the principle of methodological individualism as follows: 'According to this principle, constituents of the social world are individual people who act more or less appropriately in the light of their dispositions and understanding of their situation. Every complex social situation, institution or event is the result of a particular configuration of individuals, their dispositions, situations, beliefs, and physical resources and environment. There may be unfinished or half-way explanations of large-scale phenomena (say, inflation) in terms of other large-scale phenomena (say, full employment); but we shall not have arrrived at rock-bottom explanations of such large-scale phenomena until we have deduced an account of them from statements about the dispositions, beliefs, resources and inter-relations of individuals. (...) methodological individualism is contrasted with sociological holism or organicism. On this latter view, social systems constitute "wholes" at least in the sense that some of their large-scale behaviour is governed by macro-laws which are essentially sociological in the sense that they are *sui generis* and not to be explained as mere regularities or tendencies resulting from the behaviour of interacting individuals.' [3]

Thus, the main tenet of methodological individualism provides that all phenomena — and the same holds for regularities as well — which consist in the fact that certain social structures possess some properties or enter into some relations [4] or that certain relations obtain between these properties and relations — can (and in the normative version, should) be explained by reference to properties and relations which belong (range over) human individuals, or by reference to regularities that obtain between these properties and relations. One epistemological consequences of that view is that the social phenomenon identified as the fact of social acceptance (primarily by the community of researchers) of a certain class of beliefs may, and should, be reduced in the explanation to the plurality of singular acts of acceptance displayed by as many individuals as are found in the relevant social group.

As I said, this view is typical for all variants of empiricism except for Marxism. Besides, it is a premise consciously adopted, or assumed implicitly, by all epistemological orientations in which the repertoire of the basic problems arises almost exclusively from the contemplation of the relation: an individual cognitive subject vs. the object of cognition. Adoption of this point of view is a distinctive mark of all epistemological positions which, like traditional empiricism, are based on methodological individualism.

There are many variants of individualistic empiricism on which we focus here as they are positions combined with methodological individualism. They can be classified in many ways, one of which is by applying the distinction: the context of discovery vs. the context of justification. One can carry this classification a step further. One could focus on the manner of explanatory reduction of the knowledge that is socially accept-ed to the knowledge that is accepted individually — a depend-ence commonly assumed by every variant of individualistic empiricism. Or one could characterize in detail the way in which an individual acquires his/her empirical knowledge — a distinction which gives rise to different characterization of that knowledge. I will limit my concern, however, to the dis-tinction connected with the opposition: the context of dis-covery vs. the context of justification.

On the grounds of individualistic epistemology, the question: How does a given subject of cognition acquire knowledge about the object of cognition? — covers, as a matter of fact, two dif-ferent problems, depending on whether this question about the cognitive process of an individual is raised in the context of discovery or in the context of justification. In the first case, the question should read: (i) *What circumstances explain the fact that certain beliefs are accepted by an individual cognitive subject?* In the second case, the question should read: (ii) *What cognitive norms and directives applied by an indi-vidual cognitive subject make the beliefs that he/she has ac-quired valid?*

Individualistic genetic empiricism provides an answer to question (i) by implying (1) that an individual's beliefs about the world comprise a group of observational findings formed as an immediate consequence of the acts of extraspection and introspection performed by the subject, and (2) that all other beliefs (with the exception of analytical beliefs, if one wants to include them) arise in the subject as a result of conscious or unconscious mental processes which are occasioned by ob-servational findings. Individualistic methodological empiricism, on the other hand, provides an answer to question (ii) by indentifying cognitive norms and directives which ensure that an individual acquires valid knowledge, and which (1) single out a certain subset in the class of genetically observational findings, (2) specify the activities that must be conducted on the beliefs that belong to this subset of observational findings in order to obtain the remaining valid beliefs (with the exception of analytic statements if these have been included).

I will now present three historically registered orientations connected with individualistic empiricism: classical positivism, logical positivism and hypotheticism.

In classical positivism, little attention is paid to the distinction between the context of discovery and the context of justification. It is done either uncritically, if the reasons for making the distinction have not been perceived, or deliberately, when it is assumed that an answer to question (i) provides also an answer to question (ii). In both cases, epistemological investigations centre around the problem expressed in question (i). This is one of the most typical traits of positivistic psychologism, typical insofar as it reduces the problem of the validity of norms (cognitive, ethical, aesthetic, artistic, etc.) to the problem of the psychological origin of its tacit use or conscious acceptance. As is well known, this psychologism has been attacked simultaneously from many angles, by proponents of Platonic approach to philosophy of mathematics, by German humanist philosophy, and especially by phenomenologists,[5] by logical positivists and by hypotheticism. It is a striking circumstance that this criticism frequently resulted in an undermining of methodological individualism — one of the main components of positivistic psychologism. This consequence was a side effect of the adoption of apriristic intuitionism, connected more or less loosely with objective idealism.

For one, J. S. Mill was a typical representative of empiricism in its classical positivistic version. Observational findings which are a starting point of all knowledge about the world arise as a result of having certain 'feelings or states of consciousness', the remaining knowledge registers '(...) the Successions and Co-existences, the Likenesses and Unlikenesses, between feelings or states of consciousness. Those relations, when considered as subsisting between other things, exist in reality only between the states of consciousness which those things, if bodies, excite, if minds, either excite or experience.' [6] The knowledge of the world is a result of the experience that an individual collects from his/her feelings and states of mind. He/she puts them together in a manner conforming to certain regularities which connect them by relations of simultaneity or succession. The question of validity of knowledge in Mill's philosophy boils down to the question of the validity of beliefs that do not immediately concern an individual's 'feelings and states of mind'. The latter are 'selfevident truths', unlike the remaining components of our knowledge: generalizations and their consequences. These 'truths not selfevident' are apprehended by induction: '(...) all discovery of truth not selfevident, consists of inductions, and the interpretation of inductions.' [7]

Psychologism of Mill's answer to the question: Which inductive generalizations leading to conclusions that state general coincidence or succession of certain classes of phenomena are valid? — is more refined that the like attitude of many other representatives of classical positivism (e.g. H. Spencer). Although Mill says that 'We conclude from known instances to unknown by the impulse of the generalizing propensity',[8] but this does not mean that it is impossible to evaluate the validity of those spontaneous generalizations produced by psychological associative regularities in a different manner than by noting pairs of perceptions incompatible with these generalizations. Certain principles are proposed which should govern induction although these principles are themselves inductive conclusions, and therefore fallible. Mill's psychologism manifest in this approach to the problem of justification of 'not selfevident' truths cannot be identified with a simple proposition that all such statements are valid because they are generated by associative psychological regularities, but can be traced back to the claim that the criterion of valid justification of these statements is to be found in the principles which have an associative psychological origin. These principles are therefore inductive conclusions drawn from premises which describe the structure of individual acts of induction.

Psychological treatment of the context of justification, characteristic for classical positivism, can manifest itself in two ways that I have already enumerated: either by reduction of the problem of validity of directives of justification to the problem of psychological origin of the tacit application of these directives, or by its reduction to the problem of psychological origin of their conscious acceptance, leading in turn to their conscious application. As we have seen, Mill's position represents the second variant of positivistic psychologism.

For logical positivism, and here it differs from the classical version of positivism, it is typical to separate problems connected with the context of discovery, which is again treated in psychological terms, from problems of the context of justification. Epistemology is entrusted with the study of problems connected with justification, while psychology is put in charge of the analysis of the context of discovery. All this is done on the assumption that the results of the psychological analysis of the context of discovery have no bearing on epistemological research. The latter is understood as a repertoire of research procedures whose end result is a formulation of statements about the syntax or — in a later period under the influence of Tarski — the semantic properties of the language of science. 'The psychological questions — says Carnap — (...) concern the procedure of knowledge, that is, the mental events by which we come to know something. If we surrender these questions to the psychologist for his empirical investigation, there remains the logical analysis of knowledge, or more precisely, the logical analysis of the examination and verification of assertions, because knowledge consists of positively verified assertions. (...) the logical analysis of verification is the syntactical analysis of those transformation rules which determine the deduction of observation sentences. Hence epistemology —

after elimination of its metaphysical and psychological elements — is a part of syntax.'[9]

Considering the fact that Carnap's 'transformation rules'
are always based on logically true statements (analytic sentences), an answer to question (ii) about empiricism in logical
positivism (question (i) cannot even arise within that theory)
can be briefly outlined as follows: cognitively valid are all
'protocol' statements that are directly based on observation
(either extraspection or introspection), plus statements which
logically follow from non-contradictory conjunctions of the
former (verification in the strict sense), plus those that, with
the aid of certain already accepted 'protocol' statements, logically imply other already accepted 'protocol' statements (con-
firmation), plus sentences accepted on the basis of terminological conventions and their logical consequences (analytic sentences).

Let us note first of all that methodological individualism has
been rarefied in logical positivism as compared with methodological individualism of classical positivism, which is a consequence of the anti-psychological tendency, but this tendency
has not been carried on to the end.

The perceptible softening of the individualistic slant in
logical positivism is manifest in the fact that logical positivism recognizes as cognitively valid also such statements which
are not based on cognitive findings of an individual subject —
analytic sentences based on conventions or definitions of terms.
This conception has often been criticized (with good reason) for
lack of clarity (e.g. by W. O. Quine), but regardless of its
correctness, it is obvious that conventions must have a social
character: they are a result of a (tacit) agreement about the
right way of understanding certain utterances, and it is not
open to modification by individuals. This is not, of course, a
serious weakening of the assumptions of methodological individualism. Statements based on terminological conventions, i.e.,
analytic sentences, do not contain any genuine knowledge
about the world, they only help to order it, or to 'transform'

it in certain ways. Genuine knowledge is contained only in 'protocol' statements and their transformations performed with the use of analytic sentences. If we now ask the question: Why only 'protocol' statements and their transformations contain genuine knowledge of the world? — the latent psychologism of logical positivism, ostensibly rejected, leaps to the eye. For what else, if not the psychological context of discovery, gives 'protocol' statements their special epistemological role of the absolute foundation of cognition?

But Popper's hypotheticism represents a radical breake with psychologism. As he puts it: '(...) statements can be justified only by statements. The demand that all statements are to be logically justified (...) is therefore bound to lead to an infinite regress. Now, if we wish to avoid the danger of dogmatism as well as an infinite regress, then it seems as if we could only have recourse to psychologism, i.e. the doctrine that statements can be justified not only by statements but also by perceptual experience.' [10] But no statements pretending to the role of a component of scientific knowledge can be justified by an act of perception: '(...) we can utter no scientific statement that does not go far beyond what can be known with certainty "on the basis of immediate experience". (...) Every description uses universal names (or symbols or ideas); every statemenet has the character of a theory, of a hypothesis.' [11] These properties do not possibly belong to descriptions of perceptive findings, but they belong to the sphere of 'knowledge about ourselves' which lies beyond the compass of scientific, 'objective knowledge.' Consequently, although '(...) only observation can give us "knowledge concerning facts" (...), knowledge of ours does not justify or establish the truth of any statement.' [12] Observational statements (interpreted by Popper as base statements) are accepted by researchers because of their genetic connection with perceptive findings, but for science these findings are worthless as justification. For the time being, they may or may not make part of science, the question is decided entirely by their intersubjective acceptance or the lack thereof.

The strongest opposition to this contention has come from logical positivism. It has been also expressed in contemporary Polish philosophical literature. It hardly comes as a surprise. If only two possibilities are taken into account for simultaneous positive qualification of the given statement: (1) that it is based on perceptual findings (directly or indirectly), or (2) that it is a consequence of arbitrarily adopted terminological conventions, then Popper's base statements must be interpreted, if they are accepted, only as instances that belong in the second category — and therefore as statements of arbitrary character. If we reject this alternative, however, we can simply say that Popper offered no answer to the question: What is the ground of common (albeit tentative) acceptance of base statements? The causes of this acceptance are inscrutable in his philosophy; for although it grants these statements a status of elements of 'objective knowledge', at least in the sphere of 'the third world',[13] 'the third world' does not make individual researchers either accept or reject differents base statements. It is only postulated for 'the third world' that researchers should intersubjectively adjust their views on which base statements will be accepted.

It can be easily noted that this fact is closely connected with the assumptions of methodological individualism. 'The third world' does not point to any views as corresponding to base statements; but conversely, a 'resultant' of observational beliefs formed by individuals and expressed in the accepted base statements becomes always an element of 'the third world', or the Popperian correlate of social consciousness.

Hence, hypothetistic empiricism provides and answer to question (ii) by offering a characterization of the 'objective knowledge' which belongs to 'the third world'. This knowledge, always accepted partially and tentatively, emerges from application of norms and directives of 'rational criticism', which require that 'bold' hypotheses are promoted first. They must be given a form as general and precise as possible. Then they must be tested by 'severe criticism', and those which pass the demanding test of falsification constructed upon some agreed base statements and confronting the bold hypotheses with received knowledge can be validly accepted for the time being. These norms and directives of 'rational criticism' are addressed, naturally, to individual researchers. They have to be adhered to by all of them, for such is the necessary and sufficient condition of having a common cognitive result — a sole component of 'objective knowledge'.

It remains to discuss the answer of hypotheticism to question (i): How is individual knowledge formed? This problem belongs to a sphere which is called 'the second world' by Popper, and is less interesting to the proponents of hypotheticism. But we can still find some hints about the right solution of the problem in *Objective Knowledge*. The solution is anti-psychologist in two senses. First, the process of individual acquisition of knowledge — or the method of learning — is conceived in opposition to the dominant view in psychology. We learn, according to Popper, by venturing hypothetical suppositions and eliminating those that turn out to be false, that is, we learn according to elementary directives of 'rational criticism'. Association (either introspective or behaviouristic) — the psychological rendering of induction — has no role in that process. Secondly, we mostly learn what already belongs to 'the third world'. We learn by acquiring 'objective knowledge'.[14] It is a common product of the efforts of numerous researchers because knowledge, in spite of its being a joint creation, molds the knowledge of all learning individuals in a feedback process.

4.2. THE PRINCIPLE OF THE GROWTH OF SCIENCE AS THE BASIC PREMISE OF MARXIST THEORY OF COGNITION

From the point of view of Marxist theory of scientific cognition oriented towards historicism, questions (i) and (ii), which were formulated above and identified as the starting point of traditional empiricism, are badly formulated. Question (i) presupposes the existence of an unchanging class of causes operating on human individuals and making them accept various beliefs; question (ii) presupposes the existence of perennial efficiently applicable criteria of cognitive validity of concrete beliefs. Even if we suppose that these questions can be so rephrased as to make them relative to periods of history, they will still betray interests which are secondary from the point of view of Marxist epistemology. Anti-individualist orientation of Marxist epistemology requires that social rather than individual determinants of acceptance of different views are placed in the forefront of the research perspective. Such corollary would have to replace question (i), and as far as question (ii) is concerned, we will decide what to do with it further down in this essay.

An answer to the question: What are the determinants of social acceptance of various views? or the question: What determinants affect the form and growth of different forms of social consciousness? — is to be found in the first two essays of this collection. The theory of scientific cognition is primarily interested, however, in the problem of social acceptance of those classes of beliefs which are a result of research practice. What has been said about it in the previous essay, and what has been encapsulated in the principle of the growth of science, makes it possible to distinguish between two types of functional genetic determination of the fact of social acceptance of a given body of research results. I will present it now in methodological stylization, i.e., in terms of explanation.[15]

A. *The Primary Explanation.* Here the *explanandum* recognizes the fact of social acceptance of a given body of research results (especially a theory) *T*. The *explanans* declares: (1) that the body of research results *T* was obtained while correspondence with the previous state of knowledge in the relevant field has been sustained, i.e., in conformity with the principle of correspondence,[16] and (2) that the body of research results (theory) *T* performs its function objectively, i.e., it responds with sufficient adequacy to the demands of a non-scientific social practice in the scope of codification and deductive systematization of the immediate predictive elements of the subjective contexts of the relevant types of that practice.

I would like to discuss the explanation which is secondary with respect to explanation A at present, but I can do so only after I have characterized an intermediate link between them — viz., the auxiliary explanation.

A'. *The Auxiliary Explanation.* The *explanandum* states the fact of social acceptance of a set of cognitive norms and directives *M*, while the *explanans* declares: (1) that the set *M* has been obtained while ensuring correspondence to the previous state of methodological consciousness (of a given) research practice, and (2) that set *M* performs its function objectively, i.e., responds with sufficient adequacy to the demands of the relevant (field of) research practice, which means that it is 'guided' by a body of research results *T* which have been socially accepted in the previous period, and that it is capable of creating a blueprint for a new variant (of a given branch) of research practice which responds to a specific demand of a non-scientific practice.

B. *The Secondary Explanation.* The *explanandum* states the fact of social acceptance of a body of research results (theory) *T*, while the *explanans* (1) characterizes the socially accepted set of cognitive norms and directives — in manner specified by the auxiliary explanation, and (2) declares that the body of research results *T* meets the criteria of validity established by *M*.

As it can be noted, the secondary explanation is possible by virtue of the fact that research practice (like any other type of social practice) is characterized by relative autonomy. This explanation is available to be sure only if socially accepted research results have been obtained in conformance with the received methodological-theoretical humanist coefficient.[17] To pursue my quest for the principal traits of the empiricism inherent in Marxist theory of scientific cognition, I will concentrate at present on explanation A — due to its primary character.

Let us point out, first of all, that both clauses (1) and (2) included in the characterization of the *explanans* of primary explanation correspond to conditions (1) and (2) of the principle of the growth of science formulated in the previous essay. For if we grant that the principle of growth formulates a framework developmental regularity of a research practice, we must consequently acknowledge that the socially accepted successive bodies of results which arise in that framework depend on the determinants that govern that regularity. Let us have a closer look at these determinants, starting from the functional determinant (clause (2) of the characterization of the *explanans* and condition (2) of the principle of the growth of science).

According to my earlier argument, the objective function of a given type of research practice consists in codification and deductive systematization of the predictive elements of the immediate subjective social context of the relevant type of non-scientific social practice. This general formulation has to be further developed now, if only because it is exposed to a possible objection. There is no doubt that different types of non-scientific social practice have such immediate subjective contexts of predictive character that their separate elements cannot possibly be placed within the framework of scientific knowledge — if this knowledge should meet standards elaborated by the disciplines in possession of rigorously formulated theories. This objection which draws a sharp line between scientific knowledge and common sense is by no means groundless. In fact, this opposition appeared at one point in the history of scientific knowledge, and thereafter it became more and more conspicuous. I will take up this problem in a moment, but first, I would like to review the traits which characterize that set of beliefs which I called the *immediate, subjective social context of predictive character* of social practice of a given type.

This set of beliefs is a fragment of the full subjective-social context of the practice of a given type, a fragment which satisfies the following conditions: (1) it consists of beliefs of predictive character only; [18] (2) these beliefs either (i) are expressed in sentences of the form 'Always after a phenomenon of type A occurs a phenomenon of type B', or 'Always a phenomenon of type A and a phenomenon of type B occur together', where A and B denote non-empty classes (i.e., types) of token phenomena, or (ii) are expressed in sentences which logically follow from the sentences of form (i); (3) phenomena of type A are activities undertaken in definite, immediately recognizable practical (i.e., observable) circumstances, or are simply events which consist in the occurrence of these conditions; (4) phenomena of type B are observable (i.e., immediately recognizable in practice) effects of certain activities, or are observable effects of observable conditions that can accompany these activities — in the former case, the type of phenomena B is a type of the positive value by virtue of the normative component which is embedded in the full subjective-social context of a given type of practice. Particular tokens of type B are then positive, individually accepted values.

It is the intention of the foregoing description to single out from the full subjective context of a social practice of a given type the set of general beliefs (in the 'strong' sense of 'general', i.e., such beliefs whose antecedents are non-vacuously satisfied) with their logical consequences. These beliefs would represent the purely practical, socially accepted knowledge determining the form in which individuals implement socially accepted positive values. These values, carried out by individuals, not only can be materialized in the given social circumstances — so much is guaranteed by the fact that the knowledge of their implementation belongs to the functionally determined social consciousness — but are specific instances of values that are socially accepted, which means that they have been endorsed by the normative fragments of social consciousness.

I will call the contents of social practical knowledge connected with social practice in a given period in history a totality of immediate subjective social contexts of predictive character which correspond to specific types of that practice.[19] In accordance with this definition, the characterization of the objective function of the social scientific practice can be rephrased. The objective function of that practice is to *codify and systematically deduce* (by the current type of research practice and the stage of its historical development) *elements of the contents of social practical knowledge*.

As I have announced a moment ago, the description of the function of research practice will now be complemented by additional features. I will present these features using the concept of correspondence introduced in the previous essay. The relation of correspondence obtains between two sets of cognitive results (at least one of them can be a scientific theory) if the historically posterior set T' (but not necessarily temporarily posterior) has a literal reference whose sub-domain is at least as adequate approximation to practical-objective reference of the anterior set of results T as the literal objective reference of the set T. We have already distinguished two cases of correspondence between T and T': (1) T' is a simple extension of T, and then T is a fragment of T', or, to put it differently, T' explains T; (2) T' has a radically new methodological-theoretical humanist coefficient, and then it cannot simply explain T, but can supply important premises for an explanation (in functional-genetic terms) of the fact of social acceptance of T. Let us call the first kind of correspondence *explanatory correspondence* and the second kind *strict correspondence*.

Similar two cases can be distinguished with respect to the relation of a given set of research results (theory) which codify and deductively systematize the contents of social practical knowledge. It happens sometimes that this set of research results while processing the contents of social practical knowledge (i.e., through codification and deductive systematization) explains them, and sometimes strictly corresponds to them. It is also possible — and this is an intermediate case — that some elements of the social practical knowledge are simply explained by the set of research results, while others stand to the latter in the relation of strict correspondence. In effect, the objective function of the social research practice can be characterized as follows: the results obtained within a social research practice correspond explanatorily or strictly to specific elements (determined by the kind of practice and historical stage of its development) of the contents of the current social practical knowledge.

This approach helps to allay the fear, which I have already mentioned, arising from the observation that in several well developed empirical disciplines specific elements of the contents of social practical knowledge cannot be included as acceptable statements because they are incompatible with some other statements. For instance, social practical knowledge at the time of Copernicus contended that the sun moved around the earth, and so it was incompatible with the heliocentric theory. According to my approach to the objective function of research practice, the contents of social practical knowledge codified and deductively systematized by that practice need not be included among the tenets which make up the set of research results — they may be accounted for exclusively in terms of the relation of strict correspondence. In this case, all that needs to be explained is the fact that they are socially accepted. For instance, within Copernicus' theory (strengthened by additional premises) it is possible to explain the fact of social acceptance of the view that the sun revolves round the earth.

I will distinguish now two main developmental stages in every field of research practice. The objective function performs a different task in each of them. The first stage is *pre-theoretical*, and can also be called positivist, because the positivist theory of scientific cognition codifies and deductively systematizes scientific practical knowledge typically connected with this stage of the development of research practice.[20] A characteristic fact is that the results of scientific research obtained therein stand in the relation of explanatory correspondence to the contents of social practical knowledge, i.e., common practical wisdom. In other words, the research practice of that stage performs its objective function by simply explaining the registered contents of the received social practical knowledge, and does not venture beyond the limits of these contents, save for, occasionally at best, generalizing over some of their elements. The second stage is *theoretical*. Now research practice breaks out of the sphere of contents of the received social practical knowledge and constructs theories which are incompatible or logically incomparable with social practical knowledge. These theories stand in the relation of strict correspondence to particular elements of the contents of social practical knowledge. Simultaneously they predict the contents of the specialized scientific practical knowledge, and (in case when they predict successfully) explain the latter (i.e., the theories remain in the explanatory correspondence to the contents). The specifically scientific practical knowledge differs from common practical wisdom (knowledge) serving as the only empirical basis for the research practice in the pre-theoretical stage insofar as it is generated, at least in its initial stage, exclusively by the research practice that has evolved in a given field. In the practice of natural sciences employing mathematical models, the specifically scientific practical knowledge always takes the laboratory-experimental form, which is one of the most telling indicators of the fact that the practice of these sciences passed into the theoretical developmental stage at the time of Newton and Galileo. In this developmental stage, research practice functions objecively by corresponding explanatorily to the projected (predicted) elements of the contents of the specifically scientific practical knowledge. It remains at the same time in the relation of strict correspondence to the received (and this also means, to common) practical wisdom. It is a characteristic feature of this developmental stage that gradually and ever more extensively strict correspondence replaces explanatory correspondence. The intermediate links which connect the contents of the received social practical knowledge with scientific theories that function in this developmental stage usually have the form of received bodies of research results (theories) which are based on the social practical knowledge and are connected by bonds of correspondence to new theories.

The last question is connected with clause (1) in the definition of the primary explanation A. It formulates one of the two necessary conditions of social acceptance of a given set of research results (a given theory). It postulates a correspondence to the previous set of research results (a previous theory). A set of research results will be socially accepted (the theory will be socially accepted) if the following two conditions are simultaneously satisfied: (i) the condition (A.1) about correspondence to earlier results, and (ii) the condition that there be a correspondence (explanatory or strict one) between the set of results (a theory) and the received social practical knowledge, and that the former codifies and deductively systematizes the contents of the latter.

I have said already that usually a given set of research results (a theory) corresponds to the specific contents of the received social practical knowledge indirectly, through received bodies of research results (theories). Now, if establishing a correspondence between a body of research results (a theory) and an earlier body of results (an earlier theory) of the same kind always leads to the establishing of the relation of correspondence between the former and the contents of social practical knowledge as accounted for by the latter (theory), and simultaneously, if the last correspondence is the necessary condition for social acceptance of the currently proposed set of results (a currently proposed theory), then we can say that a necessary condition of social acceptance of a proposed body of research results (a proposed theory) stipulates that there be a relation of correspondence between the proposed results (theory) and those contents of the received social practical knowledge which have already been accounted for in the earlier scientific researches. This condition is so important that sometimes, as the history of science indicates, this condition alone has produced an initial social acceptance of a body of research results at least tentatively. As it is known, for instance, the general theory of relativity was greeted with a tentative acceptance among the physicists as soon as it was clear that there was a (strict) correspondence between it and classical physics, or alternatively, between it and the contents the social practical knowledge which had been molded by classical physics. It was discovered much later that the general theory of relativity corresponded explanatorily to certain new elements in the contents of the specifically scientific current social practical knowledge, unaccounted for by classical physics.

And yet, how can we know that a body of research results (a theory) which is in the relation of correspondence to an earlier body of research results (an earlier theory) enters also in the relation of correspondence to the contents of social practical knowledge connected with the earlier body of results (the earlier theory)? As a matter of fact, we know it, because the relation of correspondence between a body of research results (theory) T' and an earlier body of results of the same kind T consists in the fact, as it should be remembered, that the literal objective reference of the body of results (a theory) T' is either an equally adequate approximation to practical-objective reference of the body of results (theory) T as the literal objective reference of T (in case explanatory correspondence) or is a more adequate approximation to practical-objective reference of T than the literal reference of T (in case of strict correspondence). Now let us note that the practical-objective reference of T is nothing else but the practical-objective reference of those contents of social practical knowledge which haven been worked out by T. Hence, the correspondence between the body of research results (theory) T' and the body of research results (theory) T is also *a fortiori* a correspondence between the former and those contents of social practical knowledge which were accounted for by T.

Now, in accordance with the assumptions of the primary explanation A, the following principle can be formulated.

Social acceptance of a given body of research results (theory) T requires that there be the following bonds between T and the social practical knowledge: (1) T must correspond (explanatorily or strictly) to those elements of the contents of social practical knowledge which have been accounted for in earlier scientific researches to which T also corresponds (this condition is a rephrasing of clause (1) in the *explanans* of the primary explanation A); (2) T must correspond explanatorily or strictly to some new elements of the contents of the social practical knowledge, and it must correspond explanatorily (this condition is a rephrasing of clause (2) in the *explanans* of the primary explanation A) to those elements of the contents of the specifically scientific practical knowledge which it projects.

It seems proper to me to call the above principle the *principle of empiricism of Marxist theory of scientific cognition*. It specifies in what relation to social practical knowledge should remain any theory belonging to empirical sciences — or more generally — in what relation to it any body of scientific research results should remain if they are expected to acquire social acceptance and make it possible (which is the same thing) that the social research practice efficiently performs its objective function.

Two considerations have to be highlighted. First, the principle of empiricism formulated above concerns social practical knowledge and not private practical knowledge of sundry individuals, which is consistent with the assumptions of methodological anti-individualism. Secondly, as it can be easily noticed, this principle heavily relies on the principle of the growth of science formulated in the previous essay (with the additional help of the formula of primary explanation A which corresponds to the principle of the growth of science). We may conclude therefore that the rephrased normative principle of empiricism, which encompasses as a *conditio sine qua non* of the validity of any empirical knowledge the two necessary conditions of the social acceptance of a given body of research results belonging to an empirical discipline, postulates implementation of such cognitive values (acquisition of valid empirical knowledge) whose proliferation conforms with the permanent developmental exigencies of social research practice. These in turn are determined by the character of the objective function performed by that practice with respect to the totality of the developing social practice.

4.3. SOME REMARKS ON THE CONCEPT OF SCIENTIFIC VALIDITY

The rephrased normative principle of empiricism which has been introduced here as a principle characterizing empiricism of Marxist theory of scientific cognition, belongs, obviously, to the normative part of that theory. It requires, to put it briefly, that any scientifically valid system of empirical knowledge should correspond to the contents of social practical knowledge accounted for by earlier scientific research (in the given discipline) and to some new elements of the contents of social practical knowledge. The *norm of* (Marxist) *empiricism* is a counterpart of the traditional solution of the same problem formulated as question (ii) in the opening part of this essay. Similarly, the principle of empiricism is a counterpart of the problem traditionally expressed as question (i). I am speaking about counterparts, and not alternative solutions of the same problems, for two reasons. First, as I said before, questions (i) and (ii) were formulated on the grounds of a conception which chose ahistoric approach to social (or rather individual) methodological consciousness and espoused methodological individualism. It is clear therefore that Marxist theory of scientific cognition cannot offer direct answers to them. It may propose answers to slightly different questions about determinants (formulated in framework terms) of social acceptance of various systems of empirical knowledge and about conditions (again formulated in framework terms) of scientific validity of these systems. Secondly, the principle and norm of empiricism, which have been formulated above, are the required answers to these questions provided by Marxist theory of scientific cognition. But their contents do not quite coincide with the run-of-the-mill answers of traditional empiricism to questions (i) and (ii). My answers give only necessary conditions of social acceptance and scientific validity of different systems of empirical knowledge. It should be so, I believe. The conditions mentioned by the principle and norm of empiricism stipulate that bonds of appropriate type should connect systems of knowledge with social practical knowledge. Additionally, these systems should meet some conditions concerning their relatedness to earlier results of scientific research, and namely those which satisfy the principle of correspondence as formulated in the previous essay. Only after all this has been ascertained, can we say that the requisite sufficient conditions of social acceptance or of scientific validity of a system of empirical knowledge have been fulfilled.

I would like to turn to a different problem now. I have said above that the concept of scientific validity of a given system of empirical knowledge is a normative equivalent of the concept of social acceptance of that system. From the point of view of assumptions which I have made elsewhere,[21] it means that the descriptive sense of these two concepts is in point of fact co-extensive, and the concepts mainly differ insofar as the former has an additional axiological component — its employment signifies that a certain state of affairs has a positive value, that a given system of knowledge is scientifically valid. This value, to be sure, is a particularly positive value of cognitive type in the eyes of Marxist theorists of scientific cognition. Marxism characterizes this value in its characteristic manner as a definite framework state of affairs which assumes various concrete forms in different historical stages of the development of research practice. However, to repeat, on the purely descriptive plane the statement that a given system of scientific knowledge is valid in such and such historical context is equivalent to the statement that the system is socially accepted in that historical context. Inevitably, this conception of scientific validity will be unacceptable especially to those who reject the tenet that Marxist theory of cognition is a historical discipline. They will say, first of all, that this tenet is a manifestation of historical cognitive relativism in a radical form. But is it really so? I will try to argue for the negative answer now.

My starting point is the belief that Marxist epistemology (called dialectical materialism if taken together with an ontology of its own) is not 'a science before science', in the sense of a discipline offering primeval (pre-scientific) norms of valid cognition whose observance allegedly quarantees true cognition (in the classical sense of the term) in all sciences at all times. Rather, every theory of scientific cognition, and this also holds for Marxism, is a double answer: a direct answer to an objective demand of a research practice which needs to be fortified with premises proving its rationalization; and an indirect answer ('in the last resort') to an objective demand of a non-scientific practice which has initiated an appropriate research practice that is in need of rationalization by the theory of scientific cognition. This is the meaning in which Marxist theory of scientific cogniton is an answer to the demand of the research practice based on historical materialism, of that practice which, in the words of Althusser, 'discovered the continent of history' and submitted it to a penetrating investigation. This fact reveals the character of Marxist theory of scientific cognition. The theory is a history (in the Marxist sense of the term) of one type of social practice and one form of social consciousness connected with it. It is a history of research practice, and especially of its social-subjective conditions manifested as a social methodological consciousness and its products represented by socially accepted systems of knowledge. As a historical discipline, Marxist theory of scientific cognition is built upon historical materialism and consequently interprets the development of the research practice as a fragment of the entirety of developmental processes whose most general treatment is to be found in the theory of historical materialism.

Marxist theory of scientific cognition, as I understand it, not only undertakes descriptive-historical studies, that is, not only reconstructs specific contents of the social methodological consciousness in different periods of time and not only explains historical changes that occur in the scope of its duration, but also formulates its own normative component. This component, however, does not represent a perennial model of an ideal research procedure independent of the results of historical investigations. On the contrary, it is a system of cognitive (methodological) norms and directives whose historical accuracy (in Engels' sense of the term) can be demonstrated by reference to the results of historical investigations. Thanks to this method, the theory of scientific cognition developed from the premises of historical materialism is capable of finding solutions to such problems that proved too difficult to other philosophical orientations. It is particularly true about the problem of criteria of validity of scientific knowledge as proposed by philosophy — a hotly debated issue in epistemology of the 19th century. Typically, the answer was formulated along these lines: if the epistemologist intends to provide an answer about criteria of validity of scientific knowledge, he must make no use of the research results of any discipline before he has specified his criteria, or else he is begging the question he has posed, i.e., commits the error of *petitio principii*. But let us assume, for the sake of discussion, that the epistemologist did find such criteria and had made no prior use of the results reached in sciences whose validity was to be established. Then the question arises whether these criteria are themselves valid. Suppose it is possible to give the positive answer on the basis of some higher level criteria which can evaluate the original criteria. Then immediately the question of validity of the higher level criteria arises, and the questioning of this sort must go on *ad infinitum*.

Obviously, the problem arises only if epistemology is interpreted as 'a science before science'. But the Marxist point of view makes no such presumptions, as I have already pointed out.

Let us note, first of all, that the concept of scientific validity can be treated in three different ways in Marxist theory of scientific cognition. When I have mentioned it at the beginning of this section, I had this sense in mind that is usually communicated by the term a *relative truth*. I will shortly have to say more about it, but let us investigate the remaining two meaning first.

The concept of scientific validity may be viewed as a concept denoting a certain value, also called truthfulness (in the classical sense), accepted by social methodological consciousness in many periods of history as one of the fundamental goals of research practice. In this sense it is a framework value because, apart from the fact that it is constituted by a state of affairs in which a certain statement or a set of statements conforms to reality, nothing more — if we review various historical 'incarnations' of the term — can be said about it. To be sure, each historical period in the development of social methodological consciousness made this concept more precise for its own purposes (and each in its own manner) with the help of currently accepted methodological directives which characterize the method of obtaining valid scientific results (true or approximately true). It is clear that this social-subjective concept of scientific validity — or to be more precise, its current historical 'incarnation' — is always equipped with a specific social (methodological) humanist coefficient, so that the criterion of its proper employment to historical research does not have a normative-epistemological character. As historians, we make use of it similarly to the way we use, say, the concept of a work of art, which is relativized to the norms and rules of artistic-aesthetic social consciousness in the given period of time, or similarly to the way we use the concept of a sentence of a given language, which is relativized to the specific stage in the study of the grammar of that language. These concepts, like the social-subjective concept of scientific validity, are naturally used in different periods of history in a different ways, determined by currently accepted criteria that for us are no more than an object of historical reconstruction and need not be an object of cognitive-axiological approval (nor disapproval).

Therefore, when we apply a given historical variant of the social-subjective concept of validity to an appropriate system of knowledge, we can ascertain only that the system satisfies conditions of scientific validity postulated by the current social methodological consciousness. It is obvious that such a finding does not entangle us in the traditional problems of the epistemologist that have been outlined above.

However, every historical variant of the social-subjective concept of scientific validity is connected with objective conditions of social practice in the way which was extensively discussed in the previous section of this essay. Namely, if a given variant has been formed within the compass of the social methodological consciousness, then, on the assumptions that have been made above, it is a relatively adequate answer to the demand of the research practice by having rationalized it in the manner presented, and by adapting it to its objecive function. This function consists in codification and deductive systematization of specific areas of the contents of social practical knowledge through direct or indirect correspondence with them. One can say, therefore, that the acceptance of an appropriate body of research results by the contemporaneous social methodological consciousness as scientifically valid indicates that the historical variant of the social-subjective concept of scientific validity has accounted with sufficient adequacy for the objective conditions surrounding those types of practice which are associated with the social practical knowledge singled out by that variant. We have a guarantee that this account has been sufficiently adequate in the fact that the social practice which has been rationalized by the methodologically sanctioned scientific knowledge is an effective practice. This effectiveness of practice explains the fact that a particular kind of social-subjective concept of scientific validity has formed rather than any other.

Let us now turn to the second concept of scientific validity. It is more commonly used by Marxist theory of scientific cognition, and is expressed there most often by the phrase 'a relative truth'. Its range coincides with the sum of ranges of particular historical variants of the social-subjective concepts of scientific validity (truth), but it differs from the latter concept rather conspicuously by its content. This content is identical with a feature of a cognitive result tantamount to the fact that the cognitive result reconstructs objective conditions of the social practice which it rationalizes. For reasons that have already been expounded, every relative truth is covered by the range of one or another historical variant of the social-subjective concept of scientific validity, and vice-versa. By making an incidental use of the formal (semantic) concept of truth, the following observation can be made. The social-subjective scientific validity of a given body of research results (theory) is identical with the formal truth of those results in the domain which constitutes the literal (semantic) reference of those results, in the sense which was defined in the previous sections of this essay. Now, the question of relative truth hinges on the relation of literal reference of a given body of research results to the domain of objective conditions of the social practice rationalized by those results. We have called this domain the practical-objective reference of the pertinent body of research results. Thus, specifically, a given body of research results is relatively true if its semantic reference sufficiently approximates to its practical-objective reference. Simultaneously, it can be assumed that the semantic reference of a given research result sufficiently approximates to its practical-objective reference if on the grounds of historical materialism it is possible to explain the fact (to sanction methodologically the fact) that that result has been socially accepted, by showing how it rationalizes an area of social practice which is observed in the prevailing objective conditions (in the appropriate domain of its practical-objective reference).

It is easy to notice that not only the first one but also the second concept of scientific validity, i.e., the concept of relative truth, requires no normative-epistemological justification. In conformity with the assumptions I have made, the concept is founded on the thesis that the knowledge which rationalizes social effective practice always has a literal reference which is sufficiently close to the objective conditions of that practice (and that indeed makes that knowledge relatively true). Hence, also in this case the traditional trouble of the epistemologist, already discussed above, does not arise at all.

The third concept of scientific validity appears in the context of historical investigations of the sequence of relative truths, and especially when the connection between each link S' and its immediate predecessor S in the sequence is traced down. As we have established, the relation between S and S' is that of correspondence and has such property that the semantic reference of the system of knowledge S' is at least as approximate to the practical-objective reference of the system of knowledge S as the semantic reference of system S. The links of the chain of correspondence are therefore ordered in a series of elements of non-diminishing degree of relative truth. That means that, from one element to the next, the degree of relative truth either remains constant or rises.

Cases of the first type are represented by explanatory correspondence; cases of the second type by strict correspondence.

It is crucial to be aware of the fact that the thesis to the effect that successive stages in the development of scientific knowledge form a chain of correspondence is not a simple historical generalization which records a fact discovered by the historical analysis of the development of knowledge, as it has been observed so far, but is a consequence of the general characterization of the objective function performed by research practice for the benefit of all types of social practice from the moment when research practice has emerged as an autonomous part of the social division of labour. The tendency of the relative truth to remain at a constant level or to go up can be called the process of approaching the absolute scientific validity (the absolute truth). It is worth being noted that a related insight underlies a well known opinion of Engels about truth as a process.

It is evident *a priori* that absolute truth, as an attribute of the last link in the chain of progressing cognition, the link which represents full identity of a literal (semantic) reference with the practical-objective reference, is not a concept that can be effectively predicated of any historically conceivable cognitive result. Hence, the question about the criteria of its application cannot possibly arise. A question like that does arise with regard to the relation of higher degree of relative truth, which accompanies the progress of scientific validity towards absolute truth. But the criteria of the proper use to the concept which denotes this relation, viz., the criterion which identifies the bond of strict correspondence, do not pose any new problems over those which have been reviewed in connection with the predication of relative truth.

Let us observe that the concept which denotes the relation of progressive approximation to absolute truth — and not unlike the concept of relative truth, for that matter — has an axiological-cognitive sense alongside the historical-descriptive sense.

The former is expressed by the epistemological norm which provides that every new system of scientific knowledge must stand in the relation of a closer approximation to absolute truth, i.e., that it stands in the relation of strict correspondence to the received systems. It is not so, however, that particular stages in the development of knowledge constitute a chain of correspondence because this norm is perennially binding on all scientific practice; but conversely, scientific practice satisfies the principle of correspondence because it is subjected to framework developmental regularities, which I tried to pinpoint in the previous essay by formulating the principle of the growth of science. Nothing short of the recognition of the last regularity can support the norm about approximation to absolute truth.

To sum up the foregoing inquiries, I can point out that the thesis about coincidence of the ranges of the concept of social acceptance of a given system of knowledge, i.e., the concept of scientific validity, in the social-subjective sense (the first one out of three that have been discussed above), and the concept of scientific validity in the sense of relative truth, is not a thesis of cognitive relativism of some sort. Just the opposite, it presupposes a development of research practice that can also be called a cognitive progress, assuming that the latter concept embraces the same historical-descriptive connotation as the former one, and additionally possesses an axiological component.

4.4. SOME REMARKS ON CONCRETIZATION OF IDEALIZATIONAL LAWS

The foregoing inquiries supply premises for the conclusion that the normative part of Marxist theory of scientific cognition contains two kinds of methodological norms and directives. Some of them, like e.g. the directive of correspondence, are based on the recognition of certain framework regularities which govern the development of social research practice in all periods of history, being a consequence of certain framework invariants in the mode of objective functioning of that practice. Others are found only in the relatively advanced developmental stages of research practice and reveal developmental necessities typical only for these stages. I think that, for instance, the directive of abstraction, or related to it directive of concretization, have such character. Both belong to the normative part of Marxist theory of scientific cognition, and are found in the more advanced developmental stages of research practice.

I will not attempt to present these two directives in a precise way, as I believe that what can be called the procedure of abstraction (and concretization) — and what retains the insights of the founding fathers of Marxism, and better yet, what follows from the assumptions of historical materialism — cannot be precisely described before many new analyses have been completed. I will limit my treatment of abstraction to a general characterization of the term in its broad sense. This form of abstraction was known and applied, especially in mathematical-natural science, long before Marxism has been around. The procedure of abstraction in the specifically Marxist sense is a singular instance of the former; an instance characterized, first of all, by incorporation of the method of historical development of concepts, in my opinion.

The procedure of abstraction, in the broad sense of the term, consists in a mental isolation of a certain substructure from the structure, or a relational system, which underlies the domain (or its part) submitted to investigation by a research practice of a specific science. The substructure is extracted from its surroundings in order to discover regularities which characterize it, i.e., in order to identify dependencies between several types of phenomena occurring in that substructure. It differs from the initial structure at least insofar as it does not possess some of the characteristics of the latter, or its universe is a proper subset of the entire structure, or both conditions are simultaneously true. One of the simplest

examples of such abstraction is, for instance, a mental (and possibly also an experimental) isolation of a structure constituted by
an animal organism and a substructure constituted by an ali-
mentary, respiratory, circulatory, etc. system, performed in order to study the regularities which occur in the subsystem.

It was an act of abstraction in the same sense when Galileo
isolated from the relational system representing the manifold
of the physical world the substructure, which I will now describe. It comprised: the two term relation of gravity, that
means, a set of ordered pairs of elements of the form ⟨the
earth, any physical object⟩, the three term relation of the
length of the path from one spatial location to another that
a given, arbitrarily selected physical object has travelled, the
three term relation of the duration of travel made by that
arbitrarily selected physical object (the relation between two
spatial localizations and that object), the two term relation
of distance between two spatial localizations of a given physical object, the two term relation between a given physical
object and a given spatial localization (the relation of remaining in rest within the boundaries of that localization), and the
three term relation of falling of a given physical object from
a certain localization in space on a spatially localized fragment of the surface of the earth. All other relations and properties of physical objects have been neglected.

Let us note that with the exception of the last relation all
other characteristics of the isolated substructure can be expressed as magnitudes, and this means that there are functions which assign real numbers to particular cases of these
relations. These numbers provide numerical measures of the
particular cases. This is a characteristic trait of almost all
abstractions made in contemporary, mathematically oriented
natural sciences whose history begins with Galileo. It seems
to me, in fact, that abstraction created a situation when large
scale operation with magnitudes became possible, when it
was possible to describe regularities with the help of corresponding functions which map out particular cases onto numeri-

cal measures. This is not to say, of course, that particular entries in the mental characterization of isolated substructures must either be magnitudes or that they are not susceptible to abstraction. We can point to a number of examples in biological or humanistic disciplines where the procedure of abstraction has been used without simultaneous application of quantitative concepts to denote them.

To return to the example given above, the relation of gravitation corresponds to the function $g(E, x)$, where E stands for the earth and x ranges over the class of all physical objects. According to the view of Galileo (corrected at a later time by Newton), this function has the constant value for all x, amounting to 9.8 m/s² (this magnitude can be symbolized by g). The relation of the path travelled by object x from point y (spatial localization [22]) to point z is mapped out by the function $l(x, y, z)$ whose values (numerical measures to be applied to the path travelled), in case of straightforward movement, are equal to appropriate values of the function $d(y, z)$ assigned to the relation of distance between y and z. The relation of duration of travel made by object x from point y to point z is connected with function $t(x, y, z)$ which assumes values from the set of numerical measures of temporal intervals.[23] Finally, the relation of remaining of in rest which obtains between object x and point y corresponds to the function $V(x, y) = 0$.

In this isolated substructure characterized above we can consider a physical object x in a free fall from point y, which means that $V(x, y) = 0$, towards point z on the surface of the earth. According to Galileo, the regularity of its movement can be described by the function.

$$l(x,y,z) = d(y,z) = \tfrac{1}{2} g \cdot t(x,y,z)^2.$$

This inscription constitutes the consequent of the conditional which represents the law of free fall in its full form according to Galileo. Let us note, however, that the full form of Galileo's law of free fall can be presented in two ways.

First, we may decide that this law refers only to the isolated substructure described above. If so, then the antecedent will contain only the condition that x is a physical free falling object (travelling from y to z), and the adjective 'free' means that we apply this law only to the isolated substructure, freed, as it were, from any external relations that might obtain between object x and other physical objects, and especially that we neglect the relation of resistance by the surrounding air. The antecedent also contains the condition $V(x, y) = 0$.[24] Secondly, we can also apply the law of free fall to the entire structure, part of which is the substructure described above. Then the antecedent must be appropriately enlarged: we must explicitly enumerate all those conditions which counterfactually deny that in the entire structure there exist properties or obtain connections between elements of the substructure or the entire structure, which, according to our knowledge, counteract the regularity which has been postulated for the isolated structure. For instance, in the antecedent of Galileo's law of free fall we must include the caveat that the effect of air resistance on the free fall of a physical object has been neglected, because the process of gravitational falling of a physical object surrounded by air unfolds differently from the regularity described by the function representing Galileo's law. The law has been formulated for a substructure which excludes the specific relation between the falling object and the surrounding air. Consequently, this omission has to be counterfactually endorsed and explicitly incorporated in the antecedent of the law of free fall, for fear that, as far as we know, the actual properties or relations which obtain between the objects investigated will falsify the law when we refer it to the entire structure.

A law formulated as a result of application of the procedure of abstraction but projected back on the entire structure is called an *idealizational law* in the Poznań methodological circle, and the conditions included in the antecedent which declare that certain relations and properties which are absent in the isolated substructure will be excluded from the picture of the entire structure are called *idealizing conditions* or *assumptions*. These conditions, it should be noted, often refer to stepwise properties or relations, and may be expressed by predicates which denote extremal, 'minimal' modifications (specific instances) of the investigated properties or relations. If these properties or relations are magnitudes, these predicates declare that the functional magnitudes under investigation assume 'minimal', limit values, usually the value of zero. Certainly, from the point of view of the entire structure, the idealizing conditions are never actually satisfied.[25] I may add that by virtue of this happenstance, the idealizing conditions could more aptly be called the *isolating conditions*, as they caution that certain influences on the regularity being investigated have been omitted in the overall picture; those namely which are exerted by properties or relations not accounted for in the isolated substructure.

The elements of this characterization of the idealizational laws rest in large measure on the description which I have provided in the book *Z metodologicznych problemów interpretacji humanistycznej.*[26] It has been expanded herewith by including the connection between the idealizational law and the procedure of abstraction. Consequently, a further remark will be in place. The initial entire structure on which the procedure of abstraction is performed is not given in an original, intuitively-aprioristic insight. Its image has been shaped by the received system of scientific knowledge or even by an appropriate fragment of the contents of social practical knowledge (if the structure in question has not yet been made an object of scientific inquiry). When the process of abstraction has been completed, we formulate an idealizational law. Then very often we undertake another, and yet another procedure of abstraction — we isolate mentally different substructures of overlapping universes or different characteristics. In effect, when the series of abstractions has been finished, the picture of the initial structure does not necessarily have to remain intact in the form that was suggested by the received knowledge. We must mentally put together images of disparate, isolated substructures. I call this process *concretization*. Thus, the overall picture of the entire structure emerges from the process of concretization, and it may differ from the picture supplied by the initial knowledge. To put it differently, the initial knowledge may be transformed by a series of abstractions and the process of concretization. If so much is true, then the character of the conditions to be enumerated in the antecedent of an idealizational law, as idealizing or isolating conditions, cannot be determined by our initial knowledge but by knowledge resulting from concretization. Hence, each conception of the idealizational law and each concept of idealizing condition (assumption) should be relativized to the knowledge resulting from concretization.

That our knowledge changes in the process, and the picture of the 'whole' after abstractions-*cum*-concretization have been performed differs from the initial picture of the 'whole' can well be evidenced by the case of the law of free fall as formulated by Galileo. It was part of the received knowledge, largely shaped by Aristotle's physics, that the process of free fall was influenced by the 'weight' of the falling object. This belief was eliminated by Galileo and made no part of this physics.[27]

Cosequently, the formulation often encountered in publications on idealizational laws to the effect that concretization 'reinstates' 'realistic' conditions which were suspended by idealizing conditions is misleading. Frequently, what has been an idealizing conditon from the point of view of the initial knowledge turns out not to be an idealizing condition in the light of subsequent knowledge but rather a 'realistic' condition, for instance, the condition that the 'weight' of a falling object does not effect the process of free fall. And just as frequently, it becomes necessary after concretization to introduce new idealizing conditions which did not seem to be necessary from the point of view of the received knowledge.

We have reached a point here which touches upon the

central issue in this fragment of my analysis: the problem of confronting idealizational laws with experience. As must be clear by now, I have not given it a satisfactory treatment in Z *metodologicznych problemów interpretacji humanistycznej*.[28] I made an assumption there that after completion of a mental adjustment of the substructures isolated by abstraction, i.e., after the process of concretization has been finished, an automatic correction is made for the intervening influence of the properties or relations omitted in the description of the substructure. The correction describes that influence which the omitted elements have on the regularity described by the idealizational law, and the formulation of that regularity must be appropriately modified with the help of the principle of coordination — a term introduced by L. Nowak.[29] The idealizational law has also to be modified, and its consequence arrived at by concretization can now be called the factual law — again a term introduced by L. Nowak. The factual law is allegedly a kind of phenomenalist generalization to be directly confronted with observation: particular instances of logical implication which represent individual cases subsumed by the law. As we can see, this approach takes it for granted that a confrontation of an idealizational law with experience reduces to the 'reinstatement' of the 'realistic' conditions which have been replaced by idealizing (isolating) conditions in the procedure of abstraction, while both the idealizing and the 'realistic' conditions are relativized to the initial knowledge. It is assumed, in other words, that the initial knowledge comprises appropriate elements of the contents of social practical knowledge (represented by factual laws) and that we return to it after the process of concretization. The picture of the entire structure will therefore not differ in the knowledge purified by abstraction and concretization from what it has been in the received knowledge which contained elements of social practical knowledge.

I have said already that this assumption is inherently fallacious. The picture of the entire structure which we have

after a series of abstractions wrapped up with concretization have been performed can be, and usually is, radically
different from the picture provided by the initial knowledge.
This simply means, on the assumptions made in this essay,
that the resulting knowledge stands in the relation of strict
correspondence to the relevant components of the initial knowledge. Consequently, it also stands in the relation of strict correspondence to the relevant elements of social practical knowledge comprised in the received knowledge. If no earlier scientific research on the entire structure has been conducted, the
new knowledge stands in the relation of strict correspondence
only to relevant elements of the contents of social practical
knowledge. Only in case when a series of abstractions followed
by a process of concretization actually produces the same (or
possibly, more detailed) picture of the initially studied structure as the one which belonged to the contents of social practical knowledge (and possibly, received knowledge), can the
idea of conflating the idealizational law with social practical
knowledge be accepted. But such situation occurs only excep-
tionally, if at all.

Besides, I believe that from the point of view of the normative part of Marxist theory of scientific cognition the process
of abstraction-*cum*-concretization should never lead to such
results. Karl Marx, for instance, by putting science (in the
normative sense) and common practical wisdom widely apart,
undoubtedly postulated a relation of strict correspondence,
as I called it, between science and the contents of the common
practical wisdom — the latter being only capable of presenting
the 'appearance' and not the 'essence' of what there is. So he
could not conceivably accept the thesis that we use abstraction
and concretization in order to return in the end to the picture
of the world that we have found in the social practical knowledge.

If a scientific theory is conceived as a system of empirical
knowledge remaining in a strict correspondence with at least
some elements of the contents of the common practical wisdom,

but not with all of them, the methodological role of concretization can be characterized as follows. First, it is not a rule governing the operation which reconciles idealizational laws with the common, received practical wisdom in this simple way that it generally endorses incorporation of the elements of the contents of practical wisdom among factual consequences of idealizational laws. Secondly, as a result of completing the process of concretization, the researcher remains on the theoretical level. The picture of the entire structure arrived at in the process of concretization is still a theoretical-concrete picture, in specifically Marxian (or perhaps also Hegelian) sense. Thirdly, a confrontation of idealizational laws with experience unconditionally presupposes implementation of the process of concretization. But the confrontation itself is essentially something different. It proceeds mainly towards ascertaining the nature of the correspondence — which, generally speaking, must be some form of strict correspondence — between concretized theoretical knowledge and the related contents of the received social practical knowledge (possibly including a received scientific knowledge), which has provided the initial picture of the entire structure for the procedure of abstraction. It proceeds also, in accordance with the findings made earlier in this essay, to establish the relation of correspondence of either strict or explanatory kind (the latter case obtains in the context of theoretically projected, specifically scientific practical knowledge) between the new knowledge and new elements of the contents of social practical knowledge.

So much at least must be included, in my opinion, in the characterization of the process of abstraction-*cum*-concretization made from the point of view of the historical-descriptive part of Marxist theory of scientific cognition. Initiation of the procedure of abstraction appears to be possible in this perspective only after an appropriate set of elements of the contents of social practical knowledge took a definite form within the system of pre-theoretical scientific knowledge. If this has happened, the received knowledge presents a certain 'whole', on which the procedure of abstraction can be performed and which leads, after concretization has also been applied, to the creation of a new picture of that 'whole', this time being a purely theoretical picture. Naturally, this new picture of the 'whole' can serve as a starting point for performing a new round of abstracting which promotes a further development of research practice.

From the point of view of the normative part of Marxist theory of scientific cognition, the procedure of abstraction-*cum*-concretization leads to the results that represent positive cognitive values. Consequently, it is regulated by appropriate methodological norms and directives. But, as I mentioned before, these norms and directives presuppose narrower concepts of abstraction and concretization from those defined above. I will not present these specifically Marxist concepts here, because, as I said before, to do so would be immature considering the present state of research. I will only give some hints about topics that should be tackled in further investigations of these problems.

First, the procedure of abstraction-*cum*-concretization in the specifically Marxist sense of the term is directly connected with the directive of correspondence (and normally with strict correspondence). This means that a confrontation of the picture of the entire structure produced by this procedure and experience should provide an explanation of the fact of social acceptance of those elements of the contents of social practical knowledge, or alternatively, of the system of research results, which combine to make up a picture of the initial 'whole' subsequently submitted to the abstraction procedure. In other words, one is looking for an answer to the question: Why has the social practice which prevails in the objective conditions characterized by the theoretical, concrete picture of the 'whole' (and therefore exercised in the context of that 'whole') determined functionally such image of the 'whole' as is projected by the initial (pre-scientific, Marx would say) conception of the

'whole'? The fact that the founding fathers of Marxism applied the directive of strict correspondence in this form will
be patently obvious if we follow Marx' analysis in *Capital*
of the 'surface of phenomena', i.e., a picture of the capitalist
mode of production formed in the consciousness of its subjects
(its 'agents'), codified and systematized by the bourgeois vulgar
economy which contained elements of the contents of the social practical knowledge of the capitalist. Obviously, it was
not Marx' concern to derive deductively elements of capitalist
consciousness (e.g., the belief that the price of the commodity
is determined by the relation of demand to supply) from his
own theory of the capitalist mode of production and thereby
make these beliefs part of his theory. He wanted to examine
the relation of strict correspondence between these beliefs and
his own theory in order to be able to explain the fact that
these beliefs were accepted and to confirm in this way the
picture of the entire system of mechanisms which rule in capitalist society — as they were presented in *Capital*.

Secondly, concretization postulated by the normative part
of Marxist theory of scientific cognition should, in my opinion,
proceed in a way adjusted to the method of historical unfolding
of concepts applied by the founding fathers of Marxism. This
requirement indirectly characterizes to some extent the procedure of abstraction (in the specifically Marxist sense of the
term), since the result of the application of abstraction must
have such properties which will make it amenable to a further
transformation in the process of concretization guided by the
rule of historical unfolding of concepts. Such concretization
must possess certain minimal properties.

A substructure, called by Marx a 'category', is isolated
from the context of the 'whole' delineated by the current
initial knowledge (scientific knowledge comprising elements
of the contents of scientific practical knowledge, or common
practical wisdom, consisting simply of these elements) and is
localized in such historical period in which it was still a
'simple category'. It is noteworthy that this 'category' is under-

stood by Marx in two ways. Sometimes it denotes simply a substructure which is a literal reference of a given concept (e.g., the concept of commodity). Usually, however, it denotes a substructure which represents a practical-objective reference of the relevant concept, and then it is the second meaning, of the 'category' in Marx' usage. To put it together, we abstract from the 'whole' delineated by the received knowledge a 'category' in the first sense. Then we find its 'simple', practical-objective counterpart in the former, bygone period, and there we have the second sense of 'category'. The fact that a given category (in the second sense) is 'simple' means that it is not functionally subordinated to another category and as such is the most convenient object of analysis.

But the projection into the past has other, more substantial reasons. The process of historical development transforms a 'category' that is initially 'simple' (and thereby naturally isolated as it were) into an ever more 'compound' one. This means that gradually new categories are formed by the historical process which subordinate to themselves functionally the initial 'simple' category. In effect, a hierarchical functional substructure rises in which historically posterior 'categories' are functionally superior to their earlier counterparts. The historical order of development is reverse to the order of functional dependencies, and this very fact makes it possible — or better, imperative — to discover the current functional order relying on the analysis of the historical order. The functional order characterizes the specific, current substructure which contains as its part the initial 'category', isolated before. In effect, we manage to interlock the initially abstracted 'category' with a larger 'whole', i.e., we achieve a concretization (in the sense of result of an activity). The 'category' that was abstract has become concrete, in this sense of 'concrete' which Marx assigned to this term, as far as I can see. To put it differently, we ultimately recognize this 'category' (in the sense of practical-objective reference of the relevant concept) in all its functional connections with other 'categories' (and this is exactly what it means for a 'category' to be 'concrete').

This outline of the connection between the procedures of abstraction and concretization, in the specifically Marxian sense of the terms, has been derived (after some amendments) from the results of the inquiries presented in the paper 'Uwagi o metodzie dialektycznej Karola Marksa' (Some Remarks on Marx' Dialectical Method), written by A. Pałubicka and myself.[30] The paper also contains a more developed argument to support the interpretation offered above. Here I will limit myself to a few quotations from Marx, which show convincigly, I believe, that my elucidation revives with some accuracy the intuitions of the founders of Marxism about abstraction, concretization and historical unfolding of concepts.

'(...) economic system were evolved which from simple concepts, such as labour, division of labour, demand, exchange-value, advanced to categories like State, international exchange and world market. The latter is obviously the correct scientific method. The concrete concept is concrete because it is a synthesis of many definitions, thus representing the unity of diverse aspects. It appears therefore in reasoning as a summing-up, a result, and not as the starting point, although it is the real point of origin (...) of perception and imagination. The second procedure leads from abstract definitions by way of reasoning to the reproduction of the concrete situation.' [31]

As for the point of departure of the method of historical unfolding of concepts — 'simple categories' '(...) represent relations or conditions which may reflect the immature concrete situation without as yet positing the more complex relation or condition which is conceptually expressed in the more concrete category; on the other hand, the same category may be retained as a subordinate relation in more developed concrete circumstances. Money may exist and has existed in historical time before capital, banks, wage-labour, etc. came into being.

In this respect, it can be said, therefore, that the simpler category expresses relations predominating in an immature entity or subordinate relations in a more advanced entity; relations which already existed historically before the entity and developed the aspects expressed in a more concrete category. The procedure of abstract reasoning which advances from the simplest to more complex concepts to that extent conforms to actual historical development.' [32] However, 'It would be inexpedient and wrong (...) to present the economic categories successively in the order in which they have played the dominant role in history. On the contrary, their order of succession is determined by their mutual relation in modern bourgeois society and this is quite the reverse or what appears to be natural to them or in accordance with the sequence of historical development.' [33]

NOTES

[1] This polemic not only happened, but was often conducted with some fire. It leaps to the eye at a superficial reading of Marxist texts. This impression is probably responsible for the fact that many commentators of Marxist texts cannot see the research procedure for the outbursts of polemic arguments.

[2] K. R. Popper, *The Open Society and Its Enemies*, London, 1947, Vol. 2, p. 87.

[3] J. W. N. Watkins, 'Historical Explanation in the Social Sciences', in: *Theories of History*, ed. P. Gardiner, London, 1959, p. 505.

[4] The phrase 'some objects of type A enter in a relation R' is another way of saying: 'the product of the product $A \times A$ and relation R is non-empty'.

[5] By 'German humanist philosophy' I mean all idealist conceptions created in Germany in the second half of the 19th century and in this century which in large measure concentrated on the problem of the specific methodological character of the humanities and solved it in an anti-naturalist spirit. Certainly, the problems that were studied on those occasions were much more extensive and concerned a number of issues that are traditionally included among basic philosophical problems. By way of example, I may point to the school of Heidelberg or phenomenology. The last school, as is well known, vigorously attacked psychologism, not only in epistemology of the humanities, but generally,

in every form of epistemology. For instance, R. Ingarden refers to classical positivist epistemology as 'psychophysiological theory of cognition' and declares that 'all that empiristic-psychological theory of sense data together with their causal genetic explanation has been rejected by phenomenology'. This decision is justified by saying that the manner in which observational beliefs arise has nothing to do with the validity of these beliefs. The psychological context of discovery has no epistemological significance because by deriving observational beliefs from 'physical processes with the use of causal-physiological means', one takes for granted the validity of knowledge about the physical world, although this knowledge is based on observational beliefs whose validity stands in need of examination. Cf. R. Ingarden, *Wstęp do fenomenologii Husserla (Introduction to Husserl's Phenomenology)*, translated from German into Polish by A. Półtawski, Warszawa, 1974, pp. 68–69. The text has been published only in Polish and Norwegian. The original, German version of the lectures has not been published so far.

⁶ J. S. Mill, *A System of Logic, Rationative and Inductive*, London, 1865, Vol. 1, p. 83.

⁷ *Ibid.*, p. 315. 'Interpretation of induction', or, more specifically, of its general conclusion, is, according to Mill, a deductive reasoning that uses the general conclusion as one of the premises, or a reasoning which transforms this conclusion into a directive of inference.

⁸ *Ibid.*, p. 228.

⁹ R. Carnap, *Philosophy as Logical Syntax*, London, 1935, p. 83.

¹⁰ K. R. Popper, *The Logic of Scientific Discovery*, London, 1959, pp. 93–94.

¹¹ *Ibid.*, pp. 94–95.

¹² *Ibid.*, p. 98.

¹³ Cf. previous essay.

¹⁴ In connection with this point of view, the following statement by Popper is characteristic: 'I suggest that one day we will have to revolutionize psychology by looking at the human mind as an organ for interacting with the objects of the third world; for understanding them, contributing to them, participating in them ; and for bringing them to bear on the first world.' K. R. Popper, 'On the Theory of the Objective Mind', in: K. R. Popper, *Objective Knowledge*, London, 1972, p. 156.

¹⁵ This time it will be more convenient to treat the *explanans* and the *explanandum* as linguistic expressions rather than states of affairs described by them.

¹⁶ Cf. previous essay.

¹⁷ Cf. previous essay.

¹⁸ Cf. previous essay.

[19] Social practical knowledge conceived along these lines is a result of an important modification of the earlier version of the same concept, expounded primarily in the article 'Kontekst odkrycia — kontekst uzasadniania' (Context of Discovery — Context of Justification), *Studia Metodologiczne*, No. 12, where social practical knowledge was exclusively connected by me with the basic type of social practice. The modification makes use of the concept of social practical knowledge proposed by A. Pałubicka in her article 'Praktyka — doświadczenie społeczne — nauka' (Practice — Social Practical Knowledge — Science), *Studia Metodologiczne*, 1976, No. 14. This conception has the advantage of making provisions for associating the humanities (and especially the theory of scientific cognition) with various domains of social practical knowledge. Besides, it makes it possible to explain certain specific traits of the traditional versions of empiricism by connecting them with implementation of individually accepted values. I will have more to say about these problems further down in this essay.

[20] The thesis about different developmental stages or research practice was put forward by A. Pałubicka in her articles 'Nauka a doświadczenie społeczne' (Science and Social Practical Knowledge), *Nurt*, 1974, No. 3, and 'Teoria naukowa a doświadczenie społeczne' (Scientific Theory and Social Practical Knowledge), *Nurt*, 1974, No. 4. A more extensive elaboration of this thesis plus arguments to the effect that positivist epistemology is a verbalization of the social methodological consciousness of the pre-theoretical period in the development of science is to be found in her doctoral dissertation, 'Praktyka — doświadczenie społeczne — nauka', cited above.

[21] Cf. J. Kmita, 'O niejednorodności aksjologicznej predykatów oceniających' (On Axiological Heterogeneity of Evaluative Predicates), *Studia Semiotyczne*, Vol. 4.

[22] I will replace the phrase 'spatial localization' with a simpler term, 'point', without giving the latter the normal meaning it has when it designates a material or a geometrical point.

[23] Presentation of the function which represents specific magnitudes will look differently if we put it in the form adapted to the differential version of Galileo's law.

[24] I should probably add a condition about 'sufficiently small' distance of the object which is to fall on the surface of the earth. It is an open question, however, in what form Galileo assumed this condition.

[25] One can take into account the *quasi-idealizing conditions*, to use L. Nowak's terminology. These conditions are sometimes satisfied, but in exceptional cases at best.

164   4. MARXIST EPISTEMOLOGY AND EMPIRICISM

Cf. J. Kmita, Z metodologicznych problemów interpretacji humanistycznej (Methodological Problems in Humanistic Interpretation), Warszawa, 1971, pp. 183–193.

This example must not be left without a comment. When we abstract, i.e., when we isolate a certain substructure from a 'whole', we take into account only some characteristics of the 'whole', unless, what is a purely academic possibility, abstraction consists in reducing the universe to the 'whole'. But the characterizing of the substructure can contain traits or relations, which the initial 'whole' did not posses. In the case of abstraction made by Galileo, this 'additional' relation was, e.g., the relation of gravity.

Cf. J. Kmita, cited above, pp. 191–193.

Cf. e.g. L. Nowak, Zasady marksistowskiej filozofii nauki (Principles of Marxist Philosophy of Science), Warszawa, 1974. In Z metodologicznych problemów interpretacji humanistycznej, I use the term 'the principle of applying an idealizational law'.

A. Pałubicka and J. Kmita, 'Uwagi o metodzie dialektycznej Marksa' (Some Remarks of Marx' Dialectical Method), Studia Filozoficzne, 1973, No. 11/12. The correction comes from my assumption, explicated in Introduction, that society, i.e., social practice together with its objective conditions, is a diachronic, hierarchic functional structure due to its ability to reproduce its conditions and create new conditions. Consequently, I interpret Marx' 'subordination' as functional subordination.

K. Marx, 'Introduction to the Critique of Political Economy', in: A Contribution to the Critique of Political Economy, Moscow, 1977, p. 206.

Ibid., pp. 207, 208.

Ibid., p. 213.

ADAPTATION AS THE OPPOSITION
TO CORRESPONDENCE

5.1. *WELTANSCHAUUNG* VERSUS SOCIAL PRACTICAL
KNOWLEDGE

The following are the most permanent intuitions, as it seems, pertaining to the meaning to the term *Weltanschauung*:

1. *Weltanschauung* is a set of beliefs, predominantly general, which do not describe directly any singular situations.

2. At least some of these beliefs are evaluative.

3. These beliefs are a relatively stable component of the subjective context of individual or social practical activities — and consequently, it is possible to distinguish between individual *Weltanschauung* and social *Weltanschauung*.

Attempts to make the concept of *Weltanschauung* more precise, especially by characterization of the contents of the beliefs that are encompassed within it, have been made rather frequently. It has been proposed, for instance, that these beliefs contain a certain most general picture of the world of ontological character. Or it has been proposed that a *Weltanschauung* must contain some general epistemological assumptions. I doubt, however, that in this way it is possible to form a concept of *Weltanschauung* that covers all sets of beliefs intuitively identified as *Weltanschauungen*. The domains to which (semantically) refer various beliefs which make part of one or another concrete outlook are so different that one may hope at best to single out with their help various types of *Wel-*

tanschauung systems, while any project to find invariants which characterize all *Weltanschauungen* is doomed to failure before its inception. To justify this view, it is enough to point to the differences between two domains, one, to which semantically refer beliefs counted among scientific *Weltanschauungen* and another, to which semantically refer various religious *Weltanschauungen*.

If, therefore, several versions of systems intuitively recognized as *Weltanschauungen* possess other features in common apart from those enumerated at the beginning, it would be wise to look for them, taking clue from heading 3 in the preliminary characterization of *Weltanschauung*. It appears promising to examine the type of connection between *Weltanschauungen* relativized subjectively, and social or individual practices. At first glance, it may seem that the task will be facilitated by the possibility of correlating (social) *Weltanschauungen* with specific types of social practice. Presumably, each social practice objectively generates beliefs constituting a *Weltanschauung*; and on the other hand, these beliefs function as a subjective social context, i.e., as the subjective-social regulator of the type of social practice.

But such possibility does not arise. Unlike other sets of beliefs, e.g., those which constitute science, social artistic-aesthetic consciousness, or art, religion, linguistic consciousness ('linguistic competence'), etc., a *Weltanschauung*, in the sense determined by ordinary linguistic intuitions, is not connected with a particular kind of social practice. There exists something like a research practice, an artistic practice, the practice of linguistic communication, or even religious practice, but no special practice corresponding to *Weltanschauung*. For this reason, I would no count a *Weltanschauung* among forms of social consciousness, although some authors take the opposite view.

One may draw an analogy between a (social) *Weltanschauung* and social practical knowledge. As social practical knowledge consists of socially accepted beliefs — fact functionally determined by the objective demand connected with the total number of all types of social practice (and the basic practice in the first place), i.e., by regulative beliefs which control in the social-subjective way specific actions of individuals bent on achieving their goals — similarly, a (social) *Weltanschauung* regulates in a like manner practical activities undertaken by individuals but belonging at the same time to a social practice of one kind or another. The analogy holds also in the sense that as social practical knowledge can be contrasted with individual practical knowledge (which consists of certain beliefs accepted by an individual regardless of the fact whether they are accepted by society), so social *Weltanschauung* can be contrasted with individual *Weltanschauung*. I must caution the reader now, that when I speak of *Weltanschauung* without any qualification, I will have social *Weltanschauung* in mind. I will use an appropriate qualification if I want to say something about an individual *Weltanschauung*.

The analogy noted above, which couples social practical knowledge with a *Weltanschauung,* provides a convenient platform for comparisons which will reveal differences between the two. And namely, the contents of social practical knowledge are made up (primarily) of general beliefs of predictive character which are immediately usable in practical activities. In the simplest case, they have the form: 'In (practically recognizable) circumstances C_1, ..., C_n, activity A always leads to the (practically recognizable) effect E',[1] whereas the beliefs which constitute a *Weltanschauung* are connected with practical activities in a different, more roundabout way. Besides, we usually refer to them when we look for a subjective-rational justification of entire classes of practical activities — and not their elements. To make this intuitive difference more precise, I will use a simple example. A concrete capitalist has decided to replace part of his machinery with new models. His short term, immediate goal is to increase profits. He believes that by having new means of production of appropriate kind he will be able to increase the amount of commodities produced in a unit of time, and that he will be able to sell them at the current price, and thus will increase his profit. These beliefs are part of social practical knowledge of the capitalist, although we would not count them as part of his *Weltanschauung* in the ordinary sense of the term. We would agree, however, that the belief to the effect that such practice is neutral with respect to salvation — assuming that the capitalist is a religious person — is a belief which belongs to his *Weltanschauung*. So would be the belief that the same practice increases his chances of salvation as a reward from heaven for his purveyance of larger amounts of commodities to the market — an alleged social blessing.

From the last example stems one more condition to complement those already enumerated.

4. Beliefs that make part of a *Weltanschauung* fall into two categories. To the first belong those which establish superior positive values, i.e., the 'ultimate values' that the person espouses, whose implementation is not a means to reaching other values. To the second one belong those beliefs which support the view that the values at hand implemented in practice, i.e., types of meaning assigned to practical activities are either a means to achieve those superior ends or to prevent their achievement, or are neutral with respect to their achievement.

If we conceive of a *Weltanschauung* along these lines, we can say that beliefs which are part of it justify (or, respectively, condemn) in a subjective-rational mode a class of practical activities producing effect E, which is their immediately-practical sense, by showing that values of type E are instrumental in implementation of a superior value or at least do not impede its implementation (or, respectively, make it impossible).

A picture (of subjective-social character or subjective-individual character) of the world formed within a given *Weltanschauung* will be called a *Weltanschauung*-validation of practical values [2] (or, possibly, of one of their many spheres). Construction of such a picture always presupposes identification of certain superior values and deciding what kind of relationship holds between the plethora of immediately implemented practical values and the superior values. Beliefs to which these validations appeal usually belong to different forms of social consciousness, although historically known *Weltanschauungen* always place one of them in the forefront, and draw their arguments about superior values or about the relationships between the senses of practical activities and these values from one kind of social consciousness. Consequently, we observe a domination of magical, religious, artistic or scientific thinking.

5.2. THE PRINCIPLE OF ADAPTATION

The concept of *Weltanschauung*-validation of practical values is very useful in the attempt to contrast science and arts as two different forms of social consciousness, or to put it in plain terms, in an attempt to compare research practice and artistic practice as different types of social practice subjectively regulated by these two forms of social consciousness. As I have made it clear in previous essays, in spite of historical variability of specific properties of the social research practice, it is illuminating to realize what properties it possesses in all historical variants which legitimize extracting it from the manifold of all social practice, and permit to set science apart from other forms of social consciousness. These characteristic traits belong to scientific practice due to the fact that it serves a specific objective framework function for the benefit of non-scientific practice, and primarily, for the benefit of the basic practice. This function is crucial because it gives scientific practice a place in social division of labour. Obviously, only a framework characterization of the objective function of science is possible because its concrete form is historically variable.

I have assumed in the previous essays that the objective framework function of social research practice can be defined as sustaining the relation of (explanatory or strict) correspondence with the contents of social practical knowledge by codifying and deductively systematizing these contents. Motivated by reasons similar to those which were formulated on the occasion of the discussion of research practice, I would like to propose now a certain characterization of the objective function of artistic practice. If this hypothetic characterization is correct, it explains some properties (seen as framework properties) of this practice. Let us assume, therefore, that this function comprises a *Weltanschauung*-validation of (a given sphere of) practical values. But although the objective function of social scientific practice is performed only by that practice, *Weltanschauung*-validation of practical values is also done outside arts, e.g., in magic, religion, or even science. I will not dwell on the differences between various types of validation connected with various forms of social consciousness. It is not necessary to review this extremely complex issue to be able to identify essential oppositions between research practice (science) and artistic practice (arts).

Assuming that the objective function of artistic practice has been correctly characterized above, the main opposition is this: on the one hand, scientific practice functions for the benefit of non-scientific types of social practice by codification and forecasts, which means in other terms, through the explanatory or strict correspondence, of the immediate subjective predictive [3] context of that practice, i.e., by codification and deductive systematization of socially accepted predictive knowledge applied in implementation of individual objectives; on the other hand, artistic practice, in its objective framework function, makes use of entirely different 'thought material' by highlighting the relation between immediate-practical objectives and superior values, and by ascribing a complex, axiological subjective-social (and also individual) motivation to practical activities which make part of artistic practice. This *Weltanschauung*, subjective-social, axiological motivation often played an important role in history — also in relation to basic social practice, which needs to be emphasized, as its service to other types of social practice, especially to the legal-political practice, is more visible. It will be enough to mention in this context the role of religious *Weltanschauungen* of Protestants, primarily Lutherans and Calvinists, in the formation of the commercial prerequisites of the capitalist system of production.

Theoretically it is possible to distinguish two kinds of *Weltanschauung*-validation of practical values by the artistic practice. In the first case, validation is 'dictated' by the received *Weltanschauung* system, e.g., a religious one: in the second case, validation 'guides' or proposes a new, and to some degree also novel, system of *Weltanschauung*. This is analogous to a research practice. Social practical knowledge whose contents are elaborated by a research practice is either available as received common practical wisdom or as a specifically scientific, laboratory-experimental knowledge. Possibly the analogy goes further — in the development of every domain of research practice, two stages can be distinguished: pretheoretical-and-registering versus theoretical-and-designing. Possibly similar developmental stages can be found in artistic practice. The question must be studied by historians, however, before it has been definitely decided.

Evidently, objectivity of the functions which are now discussed — *Weltanschauung*-validation of practical values — is only manifested through objective-practical effects of activities motivated by *Weltanschauung* social-subjective factors. These potential, objective global effects of social practice, recognizable also as its demand, are subjectively represented by socially accepted *Weltanschauung* superior values, although this representation is more indirect that the representation of values directly implemented in social practice and controlled subjectively by social practical knowledge. The latter values, immediate subjective representations of the objective demand, are an intermediate link, which guarantee the existence of functional determination of the socially accepted *Wel-*

tanschauung systems, and in particular, of social acceptance of historically determined methods of *Weltanschauung*-validation of practical values in the field of artistic practice. This determination selects the most efficient axiological motivation with respect to objective demand.

Let us note, however, that the efficiency of *Weltanschauung*-validation depend, but indirectly, on the efficiency of beliefs which constitute social practical wisdom or on the effeciency of beliefs representing the results of a scientific practice. In the latter case, the efficiency, which can be called predictive efficiency, signifies sufficient adequacy of knowledge with respect to objective conditions of practice. This adequacy is next followed by implementation of values which represent outcomes of activities inspired by that knowledge. In the former case, the efficiency, that can be called axiological, pertains to direct practical values by virtue of their being instrumental in implementation of superior values, and indirectly to those superior values. Efficiency of *Weltanschauung*-validation is a measure of implementation of immediate practical values, and indirectly, of implementation of the superior values connected by the *Weltanschauung* with practical values. The last circumstance provides a possibility of controlling implementation of the superior values, a chance of practical insurance that they will be implemented. A *Weltanschauung* belief to the efect that implementation of a particular immediately practical value leads to implementation of a certain superior value belongs either to social practical wisdom or to science, and may additionally possess predictive efficiency which is a proper feature of beliefs which belong to these spheres of social consciousness. But the connection is typical for scientific *Weltanschauungen*, and certainly cannot be extrapolated on all *Weltanschauungen*, on all kinds of *Weltanschauung*-validations. It will not be found, for instance, in validations of magical or religious beliefs. Therefore, although every *Weltanschauung*-validation of practical values always functions objectively as long as it has been socially accepted, the connection between the immediately practical values and superior values that it points to may have a purely subjective character ('ideological' in the negative sense of the term as frequently used by the founders of Marxism).

Turning to the arts now, when I use this term, I will always have in mind, as it has been already announced, a specific form of social consciousness, namely the artistic-aesthetic social consciousness, or, speaking generally, a set of socially accepted beliefs that identify values to be implemented in artistic practice plus those that specify rules to be observed in implementation of these values. This consciousness is functionally subordinated to social artistic practice and serves as its subjective-social regulator. As this practice is normally called 'arts',[4] I must reiterate my decision that I will use this term only to refer to the socially accepted body of beliefs. As science, also this set of beliefs which I call arts, is characterized by a certain internal structure. To reveal this structure, which extends over several developmental stages of arts, two concepts are very useful, those of 'artistic value' and 'aestetic value'.

In tune with intuitions that are normally associated with these terms, I will assume that artistic value is a means of implementation of the aesthetic value. To be thoroughly consistent, one should say that the fundamental measure of the artistic value incorporated in a work of art is the degree to which it implements aesthetic values potentially related to it. This implementation has to be understood properly. These two concepts of value, like any concept of value for that matter, have a subjective character in this (non-psychological) sense that they denote such states of affairs, which, by virtue of socially (or individually accepted hierarchy of values, are preferred over axiologically neutral states of affairs.[5] Speaking more precisely, these states of affairs are positive values (with respect to the relevant hierarchy of values), although the concept of value encompasses also negative values, i.e., states of affairs less preferred in the hierarchy of values than axio-

logically neutral states of affairs. Not to complicate my terminology excessively, I will only speak of positive values,
and for brevity will refer to them as 'values', as in fact I have
been doing so far. Now returning to the question of subjective
character of the concepts of artistic and aesthetic values, implementation of any one of them in a work of art is tantamount
to the fact that a state of affairs is connected with this work
of art that has the following characteristics: (1) it is recognizable by artistic-aesthetic social consciousness,[6] (2) it assigns a
certain property to a work of art or its part or the relation
between its parts as seen from the point of view of artistic-
aesthetic social consciousness, (3) it confers upon the work of
art a certain 'positive' rank in the evaluative artistic or aesthetic order from the point of view of the same.

In accordance with an earlier assumption (which respects
intuitions connected with the two concepts of value), implementation of aesthetic values constitutes always a subjective-
social sense of a work of art, and implementation of artistic
values is a means of implementation of that sense. Thus we
can say that the subjective, internal structure of art in one of
its domain and at a given period of history can be presented
as follows: it comprises (1) beliefs determining the aesthetic
order of values, or succinctly, determining aesthetic values,
(2) beliefs determining modes of implementation of aesthetic
values, or determining artistic values, arranged in different
orders, also with respect to their efficiency in implementation of aesthetic values or to the rank of aesthetic values.
Beliefs (1) jointly form a sphere which I call social aesthetic
consciousness ; beliefs (2) form a sphere which belongs to social
practical knowledge in my opinion, thus I call it social artistic
practical knowledge. As I have mentioned, that knowledge
can be verbalized by rules which jointly determine the
mode of implementation of different values, different aesthetic objectives. Like cognitive objectives established by social
methodological consciousness of research practice are a subjective representation of the objective demands connected

with the objective function of research practice which it is supposed to satisfy, so aesthetic objectives of arts are a subjective representation of the objective demands to be satisfied by the artistic practice through its objective function. In accordance with my earlier assumption, this objective function is performed by social artistic practice through *Weltanschauung*-validation of practical values. This means that social acceptance of a given type of aesthetic values always indicated that there is an objective demand for the type of *Weltanschauung*-validation of practical values corresponding to these aesthetic values. Hence, to each socially accepted type of aesthetic values corresponds a specific *Weltanschauung*-validation of practical values. A demand for such validation may come from various types of social practice, more or less exclusivelly controlled by the ruling class. In the final count, the demands of basic social practice are decisive because, as we have assumed, the demands of all remaining types of social practice are functionally subordinated to it.

There is a particular case which immediately clashes with the above statement, viz., the situation when a given domain of artistic practice at a given period of time responds solely, as it seems, to its own demands by implementing *Weltanschauung*-validation of its own immediately-practical objectives. This case is commonly described as the reign of the ideal of art-for-art-sake. In fact, it is a spurious exception from the principle of functional determination of the artistic practice by other types of social practice, and ultimately by basic practice. We should perceive that the situation described commonly arises when it is necessary to effect a social isolation of the artistic practice to satisfy the demands of the legal-political practice of the ruling class. This isolation limits the objective function of the artistic practice to *Weltanschauung*-validation of its own objectives. The demands of the legal-political practice of the ruling class are such that the former artistic values are transformed into aesthetic values, i.e., they become the sole objectives of the artistic practice. There is a striking analogy between artistic practice and research practice here. At the time when research objectives of a given domain of research practice cease to be a subjective representation of the objective demands of non-scientific practice, they tend to evolve in such a way as to assume the form of the formerly (axiologically) subordinate values, such as novelty, creativity, precision, etc. This is not to say that implementation of such values is unimportant for implementation of superior, more substantive cognitive values, but rather that the former have been substituted for the latter (no longer representative of any non-scientific objective demands). Deprived of external developmental stimuli research practice does not develop in substance but only puts in ever more perfect form what has already been achieved. I think that a good illustration of this situation is to be found in the humanities. One should not expect a serious, substantive growth in any branch of the humanities until a new social demand has been formed for purely theoretical research results whose impact would affect the entire social reality. Until this moment comes, the humanities remain under the spell of refinement, manifested either by progressive reformulation of the well known platitudes included in social practical wisdom with the use of ever more perfect logical-mathematical apparatus or by mass media popularization, spiced perhaps, not infrequently, by artistic ambitions.

The previously presented opposition between the objective functions of research practice (science) and artistic practice (arts) makes it possible to explain the opposition between the pattern of growth in science and in arts. Codification and deductive systematization of the contents of social practical knowledge as performed by research practice which becomes theoretically more advanced assumes a different attitude to the consequences of the accumulated 'thought material' congealed in the body of research resuls than does artistic practice. For if we grant that the social acceptance of research results is primarily determined by their predictive efficiency, in the sense accounted for above, then at no time can research practice give up on that inventory of efficient modes of action which have been previously included in the contents of systematized and codified social practical knowledge. *A fortiori,* research practice cannot give up on the contents of theories which have systematized that knowledge. Consequently, every stage of development of scientific knowledge remains in the relation of correspondence to the preceding stage. It is either strict or explanatory correspondence.[7] In the latter case, the nomothetic core of the scientific knowledge from the preceding stage is transformed to make part of the scientific knowledge of the current stage. In the former case, a new theoretical characterization is offered of a domain which, from the point of view of the current scientific knowledge, constitutes the genuine object of the past scientific knowledge, although that past knowledge defined its objective reference in other terms.

The situation is different in arts, because the efficiency of *Weltanschauung*-validation of practical values — to which corresponds the aesthetic value as the subjective-social counterpart of that validation — is measured by other factors than in science. In arts, efficiency is a measure of motivational power of different types of validation. It may seem at first glance that this efficiency can be reduced to actual availability of immediately practical values which have been *Weltanschauung*-validated. But social acceptance of *Weltanschauung*-validation administered in artistic practice through aesthetic values depends on something more. Without that additional requirement no artistic practice would be functionally compelled to stand in some relation to its own tradition. But as a matter of fact they are. In spite of the fact social acceptance of an aesthetic value ascribed to an individual work of art, and consequently social acceptance of that work of art, depends on whether it represents a validation of those immediately-practical values which currently are, or can be, implemented on social scale (and which, therefore, are a subjective-social, relatively adequate representation of the objective demands), two other factors are also important. First, artistic values are never made by starting from scratch. *Weltanschauung* forms which bind immediately practical values with superior values, and especially the patterns, which bind them together in an aesthetically acceptable way, may or may not be established in arts or in other forms of social consciousness contiguous with arts on the plane of *Weltanschauung*. In either case, they are very tenacious.[8] So much so, that new works of art cannot function socially and the aesthetic forms of *Weltanschauung*-validation cannot win motivational efficiency unless they have some affinity with the elements of tradition preserved in the sphere of social consciousness. Secondly, social artistic practical knowledge which determines artistic forms of evoking aesthetic values, and indirectly determines evoking appropriate types of *Weltanschauung*-validations, also belongs to arts (as a sphere of social consciousness). Thus one can speak of predictive efficiency, conceived along the same lines as predictive efficiency of other types of social practical knowledge, or as predictive efficiency of the results of scientific research.

Both these requirements make it incumbent upon artistic practice to cultivate a bond with its own tradition. But this connection does not have the properties of correspondence. Instead, it is a selective *adaptation*,[9] carried out as an adaptive humanist interpretation directed by current demand for different types of *Weltanschauung*-validations of practical values from the point of view of current aesthetic values which represent them. Their (axiological) efficiency, that itself is to be credited to the aesthetic tradition in some degree, decides what areas of artistic practical knowledge, as it has been known antecedently, will remain vital, and which, at least temporarily, will be eliminated. And namely, those fragments of artistic practical knowledge will remain valid which subjectively indicate methods of implementation of currently accepted aesthetic values, or such aesthetic values which can be instrumentally subordinated to current values, or (generally) can be adaptively interpreted as current. Other areas of social artistic practical knowledge, as useless, are summarily abandoned and forgotten.

In consequence, a parallel principle to the principle of correspondence [10] can be formulated, viz., the general *principle of adaptation*. It says that the indispensable condition for the social acceptance of a work of art at a given stage of the development of artistic practice in a certain field of arts is (i) implementation of at least some of those aesthetic values which are recognizable by the current aesthetic consciousness as values imbedded in the masterpieces of past epochs, or are connected with the latter by relations of instrumental subordination in the context of current, social artistic practical knowledge; (ii) continuation of all those elements of the contents of social artistic practical knowledge which are active in the creation of those relations which establish the above mentioned bonds between past and present aestehic values.

It will be easily noticed that adaptation, as a factor which determines the relationship of a given developmental stage to its predecessor, is opposed to correspondence insofar as the motivational efficiency of the *Weltanschauung*-validation is independent on the social scale from its predictive efficiency. The force of this opposition is abated by those properties of predictive efficiency which are connected with a *Weltanschauung* representative of the relevant aesthetic values. Besides, social artistic practical knowledge always has certain predictive efficiency, and inasmuch as it does, it counteracts adaptive arbitrariness.

5.3. ANTI-NATURALIST APPLICATION OF ADAPTATION TO A COGNITIVE INQUIRY IN THE HUMANITIES

From the point of view of my interpretation of the Marxist theory of scientific cognition, as expounded so far in this book, it is entirely permissible to trace affinities between earlier developmental stages of the forms of social consciousness, such as science and arts, and their later developmental stages. An example of such connection between later and earlier scientific theories will be found in the relation of correspondence statable by the principle of correspondence and postulated by the directive of correspondence. The case of the relation of adaptation is largely similar, although there is no room in that context for a methodological directive, as science and arts differ in this respect. It is important to see that these affinities should not be presented as ultimate explanatory regularities, but conversely, as consequences of the developmental regularities essential for historical materialism, i.e., as consequences of specific properties of a given form of social consciousness resulting from the objective setting of the relevant type of social practice, for which this consciousness constitutes a subjective context.

In effect, as it has already been pointed out, the internal bond of correspondence reaching from one development stage of (a branch of) science to another is determined by the relation of the scientific research practice to a non-scientific social practice. In various developmental stages of science, this relation looked differently, but the general pattern has always been the same — it is a pattern, as I said before, of functional subordination, providing a framework in which research results generate systematized, subjective predictive presumptions for practical activities which in their entirety constitute a developmental process of a (non-scientific) social practice. Obviously, scientific practice provides more than only these presumptions. The latter make but a fragment of scientific knowledge, a fragment coinciding with the scope of an appropriate sector of social practical knowledge — a sector in the sphere of social consciousness directly determined by social practice and directly motivating, on the subjective-social level, the mode of execution of the activities that make up that practice. Thus, although the effect that scientific practice brings about is a body of results, part of which only are practically functional (as contents, systematically arranged by strict or explanatory correspondence, of social practical knowledge, either in the state of formation or already well defined), this functional part is decisive (albeit as functional it never decides unequivocally) for the whole. It marks out the class of socially accepted theoretical solutions that are valid within its proper field. Hence, an inchoate or an already progressing evolution of that part, or of the corresponding contents of social practical knowledge decide about the evolution of the entire scientific knowledge. If, therefore, the objective relation of scientific practice to non-scientific social practice is such that the former theoretically systematizes the incipient or the established contents of social practical knowledge connected with the latter, then every change in the systematization requires a review of the experimental material recorded by the previous theoretical-scientific approach. It requires, in other words, that a relation of correspondence is established between the body of previous research result T, and the new body of research results T'.

The fact that another type of internal bond than correspondence can bridge different developmental stages of another form of consciousness, namely arts, is a consequence of the fact that the objective relationship of the artistic practice to other types of social practice is different from what it is in case of scientific practice. Arts perform their specifically aesthetic function of *Weltanschauung*-validation of the immediately practical values, and are thereby more conspicuously subordinated to the legal-political practice than science. In this respect, all the humanities are like arts, at least to some degree, and unlike the mathematical-natural sciences. But I need to emphasize another aspect: arts do not systematize, or predict subjective premises of practical activities that are part of the received or projected social practical knowledge (though this is what happens in the sphere of social artistic practical knowledge), but, as already indicated, they *Weltanschauung*-validate immediately practical values. Consequently, another mode of internal attachment to the accumulated 'thought material' is required. It is not so important to ensure through correspondence that patterns of efficient action accumulated in social practical knowledge and scientific knowledge remain viable, as it is to activate that material with the view to implementation of values which represent the current or incipient demand of social practice.

This analysis is supported, I believe, by the fact of a widespread acceptance of the concept of 'understanding' proffered by anti-naturalist philosophy of the humanities. 'Understanding' is nothing else but adaptation; it is a method of reinterpretation of the received works of art within a new subjective social context of artistic practice. When seen in this light, masterpieces begin to 'speak a new language', as expected to. They communicate currently accepted or acceptable aesthetic values comprised in the new *Weltanschauung*. Now they can be 'understood', because such senses have been conferred upon them and such modes of implementing them have been discerned which are familiar to the 'understanding' (i.e. adapting) subjects.

For that reason, anti-naturalist theories of humanist cognition which propose a normative conception of 'understanding' can be viewed as an attempt to transfer the internal bond characterizing the development of arts into methodology of the humanities. The principle of adaptation becomes in these theories a methodological directive of 'understanding' addressed to researchers in the humanities.

Obviously, a refutation of the norm of 'understanding' from the point of view of Marxist theory of scientific cognition is not done by merely saying that the directive of correspondence rather than the directive of 'understanding' or adaptation is valid for the humanities as much as for mathematical-natural sciences. It has to be emphasized that inasmuch as research practice of the humanities is objectively located in the social context in the proximity of scientific practice its proper functioning presupposes application of the rule of correspondence; on the other hand, to the extent that the practice of the humanities is objectively located in the vicinity of artistic practice, the proper functioning of the former implies application of the directive of adaptation.

This belief is supported by the attitude of Marxist theory of scientific cognition to the cognitive norms and directives formulated within its scope — a problem discussed by me in the previous essay. It is a fact that cognitive values generally accepted by social consciousness of the academics forbid application of the rule of adaptation of the received scientific theories, e.g., the relativistic mechanics could never be formulated by asigning appropriately modified meanings to the tenets of classical physics. As long as a given theory views its predecessor from a cognitive point of view, i.e., establishes a bond of correspondence with it, it is impossible for the earlier theory to display an ambiguity which could permit a new 'reading out' of its contents. It is one of the ideals of science to speak in a language fumigated of all forms of equivocality. But for the Marxist theory of scientific cognition the ideal of unequivocal communication in science does not stem from the proposition that unequivocality of the results of the search for truth is value absolutely binding for science, but from the regularity that has been described in this book: excessive equivocality makes practical functioning of science impossible. Truth and unequivocality are no more than relatively permanent, subjective counterparts of the objective demands of social scientific practice, and in the last resort, of the basic practice.

There is no doubt that social acceptance of the anti-naturalist program of research in the humanities which postulates 'emancipation' from the rule of the regularities statable by the principle of correspondence and subordination to the rule of adaptation objectively signifies relaxation of the functional bonds beween humanist research practice and social practical knowledge, bonds which are subjectively manifested in the acceptance of such values as truth and unequivocality. This program is accompanied by the postulate of making the humanities more responsive to practical demands which are satisfied by codification and propaganda, by promoting various forms of *Weltanschauung*-validations of the values connected primarily with legal-political practice.

The question whether the humanities should use the directive of correspondence or the anti-naturalist directive of adaptation resolves itself into the dilemma whether the humanities should practically function as science or as arts. Mindful of the fact that the fundamental evaluative category of Marxist theory of social development is the category of historical fitness, and that historically fitting is what invites or facilitates developmental changes in social reality, we can say that if it can be shown that practical functioning of the humanities as science is historically fitting, the above dilemma is as good as solved.

5.4. SOME REMARKS ON THE HUMANIST EXPLANATION OF PHENOMENA CONCERNING INDIVIDUALS

In the end, I would like to make some remarks about the assumption that I have explicitly laid out elsewhere, and namely that the tenets of historical materialism determine nothing more but methods of explanation of social phenomena, and have nothing to say about methods of explaining individual behaviour, although they are not indifferent to the last question, and indirectly legitimize some of these methods.[11]

First of all, the question comes up of drawing a boundary between social and individual phenomena. At first glance it does not seem a difficult task. Social phenomena can be understood as states of affairs described in statements which predicate certain properties or relations of social structures or substructures, or of some of their aspects (definite systems of objective conditions or their social-consciousness representations), or as states of affairs described in statements which specify connections between above mentioned properties and relations. Certainly, such phenomena have a social character. But besides and on top of these there are many phenomena which do not answer to this definition, and yet do not have a character of events concerning individuals as such. The most important in this number is the event in which a certain individual act or object is recognized as scientific achievement, a work of art or a literary masterpiece. From the point of view of common practical wisdom, the fact that, say, *Lalka* by B. Prus is a literary masterpiece, is a phenomenon concerning an individual, it is connected by natural-causal ties with individual acts of the author. But as a matter of fact, we come across a cultural object or a cultural activity of a certain kind only after this object or activity has been 'recognized' as such by social consciousness of appropriate type. And this decision is a social event. Production of an object or performace of an act are only an individual project of a cultural act or a cultural object, only part of the cause which brings them forth. The other indispensable part of that cause is an appropriate state of social consciousness which elevates these projects to the level of cultural objects of relevant type.

This conclusion gives us an important clue to be used in the solution of another problem, namely, the question of possible complementation of Marxist humanist research, i.e., of the theory which is founded on the theory of historical materialism, with conceptions which can be used as a groundwork for explanation of phenomena concerning individuals. Naturally, various psychological conceptions: psychoanalysis, neo-psychoanalysis, etc., can compete for that role. I will not focus here on different cognitive value of explanations based on such theoretical foundations. I would like to point out only that they could complement Marxist research on the condition that their theoretical presumptions do not clash with the principle of methodological anti-individualism, but apparently this condition is not acceptable even to one psychological conception. This can best be seen in the treatment of the phenomena just referred to, allegedly concerning individuals as such. They are dealt with as if they ultimately resolved in individualistic concomitances, i.e., as if they were determined by the psychological properties of the creative individual or by the properties of the individuals who appreciate the product.

A general rationale of this state of affairs is not to be found in the fact that psychology does not usually make use of the concept of social consciousness, but rather in the fact that individualistic orientation predominates in psychology. Even when psychology uses a concept that looks at first blush as a counterpart of the concept of social consciousness, e.g., Jung's concept of collective unconsciousness, it denotes only a class of identical elements of consciousness that can be intersubjectively communicated or identical contents of representations widespread among individuals.

Thus, psychology cannot complement Marxist humanist research, it can only conduct an independent, parallel research of a different type. This does not mean, of course, that psychology for whatever reasons is doomed to rely on methodological individualism. On the contrary, one can imagine psychological research which aims at a discovery of regularities characteristic of individuals who acquire the contents of anti-individualistically interpreted social consciousness. Until, however, such studies have been undertaken, humanist analysis of phenomena concerning individuals, conducted from Marxist point of view, will have to go on without cooperation with psychology. This analysis will have to adress itself to the phenomena relativized to an individual humanist coefficient (in the sense of F. Znaniecki) constituted by a creative individual, in terms which highlight the difference between that coefficient and the ideal coordinates of social consciousness. By giving precise theoretical sense to the last concept, historical materialism shows a way for humanist explanation of events concerning individuals as such, even if they lie beyond its proper scope. These phenomena, which can also be called creative phenomena, were among the topics that traditional humanist research was especially interested in. One does not see the reasons why Marxist humanist research should not be interested in them as well.

I should add that research on events concerning individuals, conducted from the point of view of methodological anti-individualism (interpreted along the lines sketched out in the Introduction, and therefore opposed to extreme anti-individualism), should have nothing to do with the naive social determinism of Hauserian type. From the point of view proposed here, individual consciousness is not a simple individualistic version of social consciousness filtered through individual personality. It is rather a 'deformed' version of the latter, made more concrete, by being in one extreme case a creative personality, and in the other extreme case a psychopathic personality. In point of fact, nothing forces an individual to respect norms and rules of social consciousness but the need to implement his/her values in a rational and efficient manner, beginning from the lowest ones, viz., the need to survive. As we go up, however, the less elementary is the value, the more probable it becomes that the individual concerned may give up on its implementation.

NOTES

[1] Cf. previous essay.

[2] The attribute 'practical' indicates that validation concerns classes of phenomena conceived of as immediate senses of practical activities.

[3] Cf. previous essay.

[4] The ambiguity concerning the term 'arts' is quite parallel to the ambiguity noted about the term 'science' and discussed in the essay 'The Controversy about the Determinants of Scientific Growth'. With the term 'science', I refer either to social research practice or to a subjective social context of that practice which comprises methodological norms and directives plus socially accepted findings.

[5] These concepts, and some other like them, have been characterized in more detail in my essay 'O niejednorodności aksjologicznej predykatów oceniających' (On Axiological Heterogeneity of Evaluative Predicates), *Studia Semiotyczne*, No. 4.

[6] We might also consider individual consciousness. But obviously, development of arts is development of social artistic-aesthetic consciousness. Although in my opinion phenomena concerning individual consciousness

play an important role in this development, I will not discuss them here, and limit my discussion to social consciousness.

⁷ Cf. two previous essays.

⁸ An intuitive grasp of this situation explains the popularity of the conception of 'archetypes' among the students of research on arts, or, more generally, students of research on culture. This conception, if it is understood according to the principles laid down by its author, C. G. Jung, makes a dare-devil attempt to explain the fact of impressive longevity of certain artistic representations ('archetypes'), lasting in some cases through millenia, and aesthetic senses which are combined with them (specific 'visions of the world' which represent appropriate types of *Weltanschauung*-validations) by biological 'racial heritage'. Most humanist, however, strip this conception of its speculative-naturalistic context, and intuitively try to give it a cultural character. On this interpretation, the concept of 'archetype' becomes a typical instance of a label-concept. If it is predicated of an object (normally an artistic one, but occasionally also o mythical or religious one), it says what has already been known about it before it was 'labelled', namely that the object represents a type that in different embodiments has been known for ages. This procedure, in spite of irrational illusions to the contrary, explains nothing. Ascription of ever more 'academic' terms to the phenomena already well known, i.e., 'labelling', has become very popular in many branches of the humanities. There would be nothing wrong about this practice from the point of view of cognitive objectives of science if the practice were not offered as a substitute for explanation. But it is, as researchers who engage in 'labelling' have the impressions that they explain something and do not feel obliged to engage in serious attempts of explanation.

⁹ This term is connected with the distinction between historical humanist interpretation and adaptive humanist interpretation. In the first case, we interpret a given activity or its effect respecting as much as possible the knowledge and the system of values of the subject of the activity in the historically accurate form. In the second case, at least some of the interpretative premises are contemporary, and they help to adapt the activity or its effect to contemporary consciousness. Cf. J. Kmita, *Z Metodologicznych problemów interpretacji humanistycznej* (*Methodological Problems in Humanistic Interpretation*), Warszawa, 1971, pp. 81, 82.

¹⁰ Cf. the essay 'The Controversy about the Determinants of Scientific Growth'.

¹¹ I would like to use this opportunity to present my opinion about the catchword-proposition of J. Ładosz: 'Historical materialism — the method-

ology of social sciences'. I believe that there is a close connection between historical materialism and Marxist methodology of sciences conceived as a theory of scientific cognition. But I see this connection differently than does J. Łodosz. First, Marxist theory of scientific cognition, as a historical discipline concerned with the development of social research practice, undoubtedly presupposes historical materialism and uses its statements as explanatory premises for its own general tenets. It also uses the concepts of historical materialism to define its own concepts. But in this sense, historical materialism is 'the methodology of all sciences', not only of social sciences. One cannot seriously contend and remain a Marxist that the growth of mathematical-natural sciences is not a fragment of social development and thus cannot be explained by historical materialism. One should say therefore that historical materialism is the methodology of all sciences. But such formula lacks precision, in the same way, as, say, the formula that biology is medicine, or that set theory is arithmetic. One can seriously claim only that Marxist theory of scientific cognition is founded, as every other humanist discipline treated à la Marx, on historical materialism. Secondly, however, I suspect that J. Ładosz opposes speculative normativism and ahistorism only in the field of ethics, but in epistemology his position is both normative and ahistorical. If so much is true, his formula should be read like this: historical materialism contains methodological norms and directives which should be applied in all social sciences (in the humanities). If this claim is understood literally, it demands more than can be delivered. One can hardly find such norms and directives in historical materialism, especially if they were to acquire the status of perennial validity for social sciences. In other words, Ładosz' formula can be given a definite, acceptable sense along these lines: the normative part of Marxist theory of scientific cognition demands that the fundamental theoretical findings of particular social sciences be explained with the help of the tenets of historical materialism. This demand is justified by that theory in a descriptive-historical manner and with the help of the concept of historical fitness. But if this is what J. Ładosz has in mind, he should avoid making misleading pronouncements.

SUBJECT INDEX